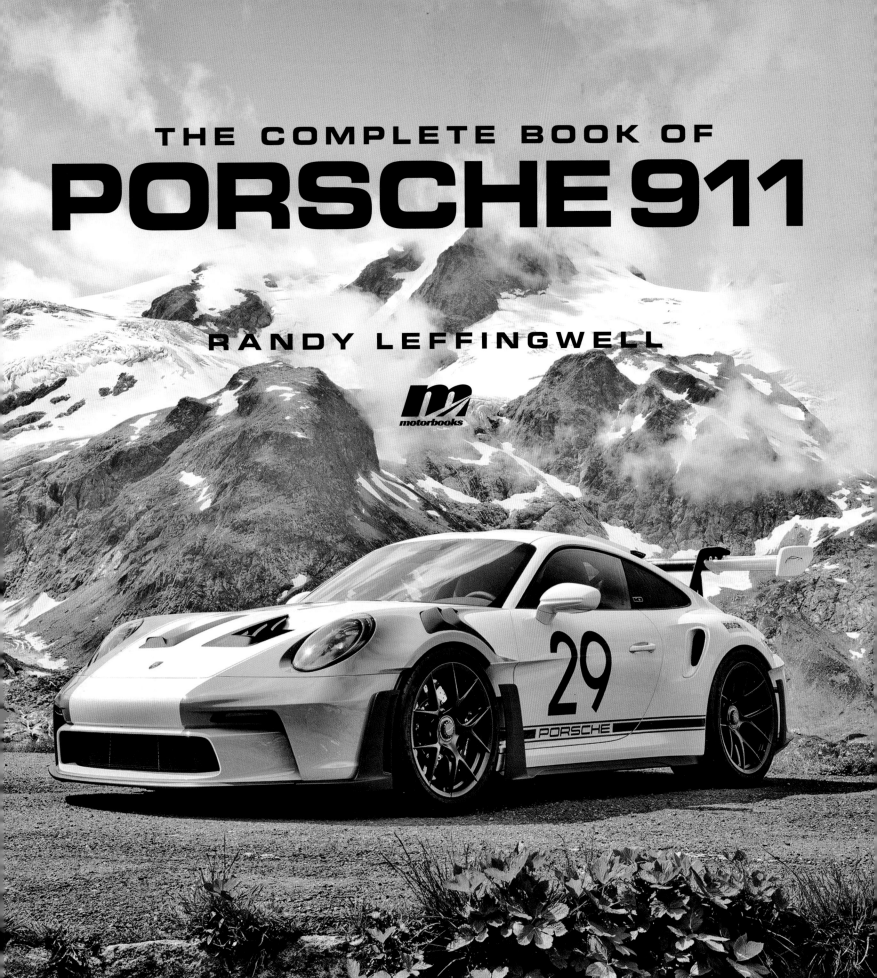

THE COMPLETE BOOK OF
PORSCHE 911

RANDY LEFFINGWELL

motorbooks

Quarto.com

© 2011, 2015, 2016, 2019, 2025 Quarto Publishing Group USA Inc.
Text © 2011, 2015, 2016, 2019, 2025 Randy Leffingwell

First Published in 2011 by Motorbooks, an imprint of The Quarto Group,
100 Cummings Center, Suite 265-D, Beverly, MA 01915, USA.
T (978) 282-9590 F (978) 283-2742

All rights reserved. No part of this book may be reproduced in any form without written permission of the copyright owners. All images in this book have been reproduced with the knowledge and prior consent of the artists concerned, and no responsibility is accepted by producer, publisher, or printer for any infringement of copyright or otherwise, arising from the contents of this publication. Every effort has been made to ensure that credits accurately comply with information supplied. We apologize for any inaccuracies that may have occurred and will resolve inaccurate or missing information in a subsequent reprinting of the book.

This publication has not been prepared, approved, or licensed by Dr. Ing. h.c. F. Porsche AG.

We recognize, further, that some words, model names, and designations mentioned herein are the property of the trademark holder. We use them for identification purposes only. This is not an official publication.

Motorbooks titles are also available at discount for retail, wholesale, promotional, and bulk purchase. For details, contact the Special Sales Manager by email at specialsales@quarto.com or by mail at The Quarto Group, Attn: Special Sales Manager, 100 Cummings Center, Suite 265-D, Beverly, MA 01915, USA.

29 28 27 26 25   1 2 3 4 5

ISBN: 978-0-7603-9388-8

Digital edition published in 2025
eISBN: 978-0-7603-9389-5

Library of Congress Cataloging-in-Publication Data available

Cover design and page layout: Silverglass
Page design: Cindy Samargia Laun

Printed in China

ENDPAPER
Front endpaper: *Porsche Corporate Archives*; Back endpaper: *Porsche Press Database*

COVER
Front cover: 1971 Porsche 911 T. *Michael Alan Ross*; Back cover: *Porsche Press Database*

Front flap: *Porsche Cars North America*; Back flap: *Dieter Landenberger*

FRONTIS
Porsche Corporate Archives

TITLE PAGE
Porsche Press Database

CONTENTS

6 — INTRODUCTION: PREDECESSORS AND PROTOTYPES, 1948–1965

30 — CHAPTER 1: THE FIRST GENERATION, 1964–1969

64 — CHAPTER 2: THE FIRST GENERATION CONTINUES, 1970–1977

102 — CHAPTER 3: THE SECOND GENERATION, 1978–1983

126 — CHAPTER 4: THE SECOND GENERATION CONTINUES, 1984–1989

160 — CHAPTER 5: THE THIRD GENERATION APPEARS, 1989–1994

194 — CHAPTER 6: THE FOURTH GENERATION, 1994–1998

220 — CHAPTER 7: WATER COOLING DEFINES THE FIFTH GENERATION, 1998–2005

254 — CHAPTER 8: THE SIXTH GENERATION, 2005–2012

292 — CHAPTER 9: THE SEVENTH GENERATION, 2012–2019

338 — CHAPTER 10: THE EIGHTH GENERATION, 2020–ON

366 — INDEX

1949 356/2 COUPE
1951 356 COUPE
1952 356 AMERICA ROADSTER
1954 356 SPEEDSTER
1955 356A CONTINENTAL COUPE
1956 356A EUROPEAN COUPE
1957 356A SPEEDSTER
1958 356A HARDTOP CABRIOLET
1959 356A-1600GS/GT CARRERA
1952 TYP 530 CLAY MODEL
1960 356B CABRIOLET
1961 356B S90 ROADSTER
1961 356B 2000 GS CARRERA 2
1962 TYP 754 T7 COUPE
1964 356C COUPE
1965 356 SC CABRIOLET

INTRODUCTION

PREDECESSORS AND PROTOTYPES 1948–1965

◀ **1961 356B Carrera 2 Cabriolet**
For several years this car was Ferry Porsche's daily driver.

A new car! For customers and automakers, this phrase signifies success, even as it hints at innumerable choices. Porsche's first decade in business brought revenues and the confidence to undertake a second-generation vehicle. Customers liked the 356s that appeared in 1948 and the 356A models the company introduced in 1956. These automobiles embodied Ferry Porsche's dream of honoring his father, Dr. Ing. h.c. Ferdinand Anton Ernst Porsche. But as Ferry related in a story he told countless times before his death in March 1998, when he looked around at the cars available to him 50 years earlier, he found nothing he liked. So with the help of his father's engineering staff, he created his own and began manufacturing them in a former sawmill in Gmünd, Austria. By the early 1950s, after returning to prewar workshops in Stuttgart, Germany, customers let him know they wanted more power, less noise, a more spacious interior with larger rear seats, and additional luggage space.

Ferry Porsche's first road cars fit the restrained attitudes prevalent in Europe following World War II. Body engineer Erwin Komenda developed beautifully proportioned coupes and open cars that slipped a function-hugging form around a driver and passenger. The 356 gave owners intimacy with style. Franz Reimspeiss's air-cooled, opposed four-cylinder engines and Leopold Schmid's stiff chassis offered lively, responsive driving experiences.

While Porsche assembled its earliest coupes in Gmünd, in the fall of 1948 the company approached Swiss coachbuilder Beutler Carrosserie to develop and manufacture convertibles. Ernst Beutler, a man of modest ambitions,

▶ Porsche assembled just 52 cars at the Gmünd facility. The coupes sold for 9,950DM, roughly $2,369, at a time when a loaf of bread cost 14 cents. This bench seat was standard equipment.

▶ (Opposite) The coupes rode on a 2,100mm (82.7-inch) wheelbase and were 3,870mm (153.4 inches) long overall. They weighed about 765 kilograms, 1,683 pounds, and were capable of nearly 140 kilometers per hour, 88 miles per hour.

▼ Porsche technicians assembled these earliest production coupes, known internally as 356/2, by hand in a former lumber mill in Gmünd, Austria, in the province of Carinthia. Porsche delivered this car on June 28, 1949, to Dr. Ernst Herschel.

stepped back from the project after completing a handful of cars. He doubted his ability to meet Porsche's future demands. For Porsche, however, the timing was auspicious; while its factory space in Stuttgart had suffered damage from Allied bombs during the war, in the early 1950s, occupation forces allowed engineers, fabricators, mechanics, and assemblers to move back into the facilities. Next door, Reutter Carrosserie was anxious to take on new work.

Porsche supported automobile shows throughout Europe and the United Kingdom with sales people and vehicles. An Austrian with substantial ambitions, Maximilian Hoffman brought Porsche cars (as well as Mercedes-Benz, Jaguars, and Volkswagens) to the United States. His vigorous personality, effective salesmanship, and obsession with detail and perfection led to success with Porsches in America. He raced the cars and encouraged others to do the same, including an equally adventurous fellow Austrian, John von Neumann, who personally drove the cars to the West Coast for his own customers. As the American market grew, Hoffman and von Neumann convinced Stuttgart management to develop new versions for their sporting and racing customers. A trim America Roadster appeared in late 1951 and 1952, followed by the rakish Speedster in 1954. Engine chief Franz Reimspeiss developed power plants of 1.1-, 1.3-, and 1.5-liter displacements for Porsche's growing product line. At the top of the performance spectrum, the new "four-cam" Carrera, designed by newcomer Ernst Fuhrmann, offered a 1.5-liter dual overhead camshaft flat four that produced 115 horsepower.

Ferdinand Porsche had died in January 1952, and his first generation of designers commingled with Ferry's younger generation of engineers. One of these newcomers was Ferry's son Ferdinand Alexander, known to family and friends as F. A. or Butzi. F. A. was among the first students at the Hochschule für Gestaltung, the upper

▲ 1951 356 1100 Coupe
The 44-horsepower 1,286cc (78.4-cubic-inch) Typ 506 engines first appeared in 1951. Manufactured in Stuttgart and assembled by Reutter, the cars had finned drum brakes with two front leading-shoes.

school for art in nearby Ulm. This institution nurtured progressive designers, emphasizing the aesthetic of everyday objects and spaces and incorporating mathematics in the design process. The faculty taught creative minds such as F. A. to "reduce ornament to a fundamental and pure form of geometry." With Germany in ruins after the war, the school pushed students to "start from new," to feel little obligation to refer to or take from the past. In his brief time at Ulm, F. A. learned to question whether old techniques applied to new products, to develop independent thoughts and ideas, and to express them confidently. He left the school in 1957, just as his father confronted the need to develop Porsche's first all-new product, a successor to the 356.

A year earlier, in 1956, the company had introduced its 356A series, with increased engine displacements of 1.3, 1.5, and 1.6 liters from Franz Reimspeiss's engineers. Neighboring Reutter body works assembled these more potent coupes, cabriolets, and speedsters.

F. A.'s father, Ferry, welcomed the new approaches his son brought home. Five years earlier, Ferry had assigned Erwin Komenda to create a four-seater, designated Project 530. Komenda, whose job title translated as "thin metal technician," had developed Porsche's sheetmetal technology and done the company's body design since 1931.

▲ Acceleration from 0 to 100 kilometers (62 miles) per hour took 22.0 seconds. The cars sold for 10,200DM, $2,429 at the time.

◀ Porsche used Volkswagen's 1,086cc (66.3-cubic-inch) air-cooled Typ 369 four-cylinder engine. It developed 40 horsepower at 4,000 rpm and drove through a four-speed transmission.

▲ (Top) **1952 Typ 540 America Roadster**
Some sources say Porsche assembled between just 16 of these 605-kilogram (1,331-pound) open cars. The 70-horsepower 1,500 S (91.5-cubic-inch) Typ 528 engines got them from 0 to 100 kilometers per hour in 10 seconds. All but one of the cars went directly to the United States for $4,600 each.

▲ (Bottom) **1954 Typ 540 Speedster**
California Porsche distributor Johnny von Neumann imagined a Porsche not only for racing, but also for cruising. Porsche responded with the Typ 540 Speedster. Equipped with the Typ 546 55-horsepower 1,500 Normal engine or the 70-horsepower 1,500 S Typ 528/2, the company charged 12,200DM for the base model ($2,905).

His old-school technical education and dozens of years of Porsche experience convinced him that strong curves imparted strength and that wide surface edges for door cuts reduced stress to thin metal.

That sufficed for their first model, but when Ferry first viewed the 530, he saw a swollen 356. Komenda's philosophy held that any new Porsche must resemble what came before. From his son, Ferry concluded that what followed need not adhere to earlier concepts. Disappointed with Komenda's effort, Ferry asked Max Hoffman to contact Count Albrecht Goertz, who had designed BMW's sleek 507. Ferry provided him a set of dimensions, and eight months later Goertz delivered Project 695. Goertz was a German living and working in New York City, and U.S. design trends influenced his creation. The 695 incorporated quad headlights, the latest rage in American car styling. His roofline descended to rear fender height, although that line ended in abruptly angled surfaces sporting three taillights per side. Seen from the side, Goertz's roof and front fender hinted at shapes to follow. However, the car just was too American for Ferry. It was around this time, in 1957, that F. A. joined the family firm, first assigned to work for overall design chief Karl Rabe, Komenda's boss.

F. A. spent time with Franz Reimspeiss, learning the engines. Then he designed his first car bodies, finishing up the cigarlike 718 Formula 2, advancing to the Typ 804 Formula One racer and then the 904 Carrera GTS coupe, done hurriedly and with little interference. From the Racing Department, he transferred to Erwin Komenda, who assigned him to help modeler Heinrich Klie reshape Albrecht Goertz's Typ 695 into something closer to what Ferry desired. Eventually,

all seven of Komenda's designers were at work on the next Porsche.

In early October 1959, F. A. and Klie completed their model on the Typ 695 platform. They adopted the 2400mm, 94.5-inch, wheelbase Komenda used for his 530. Ferry, who hoped for a sportier nature to the car, shortened the wheelbase to 2200mm, 86.625 inches. He agreed with his son that a full four-seater required a longer, flatter roof than Goertz had designed. Preferring the fastback style, because he hoped to incorporate a hatchback-type opening rear window, Ferry reassigned the rear seats to occasional use and preserved the sporting roofline. Within a year, they had a drivable prototype, fitted with a noisy flat six with dual center-mounted cams and cooled by twin fans.

Reutter contended with its neighbor's increased demand for car bodies. Porsche phased out the Speedster in late 1958 and introduced a gentrified Convertible D, manufactured by Drauz in nearby Heilbronn, when it brought out the improved 356B series later that year. (Designations got confusing; the body for the 1960 model year 356 was known internally as T-5.) To supplement the 1,600cc Normal models, the factory introduced a new, higher-performance 90-horsepower pushrod engine, the Super 90, including a series of GT models for racing.

In late 1960, Porsche changed the Convertible D designation to Roadster, still produced by Drauz. The company's most potent model, the Carrera GT, introduced the 12-volt electrical system

▲ The company offered several engines for early Speedsters, including the 1,286cc (78.5-cubic-inch) Normal with 44 horsepower and the "Super" with 60 horsepower. The base 1,488cc (90.8-cubic-inch) engine offered 55 horsepower, and the hotter S version developed 70 horsepower.

▲ 1955 356-1500 Continental Coupe
These final-year Pre-A coupes with the 44-horsepower Normal 1,286cc engine sold for $2,708 (11,400DM). A 1,488cc (91.5-cubic-inch) Typ 528/2 flat four in Super specification developed 70 horsepower and sold for $3,278 (13,800DM) at the time.

to customers. (Porsche designated these 356 bodies for the 1961 model year and later as T-6. The F. A. Porsche/Heinrich Klie four-seater prototype on the 695 platform was designated T-7.) To broaden the lineup, Porsche added a notch-roof hardtop coupe manufactured by Karmann in Osnabruck in 1961. At the same time, Ferry Porsche shifted manufacture of the Roadster to Belgium, hiring D'Ieteren Frères to produce the handsome open cars.

By late 1960, Ferry, F. A., and the engineering staff had begun working on another prototype (T-8). This was a pure two-seater built on a 2,100mm, 82.69-inch wheelbase. The front suspension system took up so much space that engineers relocated the fuel tank to the rear. Ferry encouraged the project. A working prototype, with modified 356 running gear, drove out the gates in early November 1962. Satisfied with its direction, Ferry tentatively set production start for the car for July 1963. But too many problems remained to reach that target.

As the two-seater took shape, Ferry promoted his son to head the Model Department at the beginning of 1961. Within a year, F. A. changed its name to the Styling Department. Among the engineers, a talented arrival named Helmuth Bott continued work on the car's front suspension, adopting a MacPherson strut configuration that conserved space and improved handling. This enabled Ferry to return the fuel tank to the front and enlarge it. He also reset the wheelbase to 2,211mm. Now the car was a 2+2 in what became its final form. Ferry's aerodynamicist, Josef Mickl, had worked for Ferdinand. Mickl collaborated with F. A. to fair in the new car's front bumper and to perfect the elegant sweep of the roofline.

▲ Largely intended for the U.S. markets, the "Continental" appeared more often with bench seats than individual buckets. The 356A introduced the "panoramic" windshield.

▶ In early 1955, Ford Motor Company advised Porsche that it intended to reintroduce the Lincoln Continental for 1956, so the company changed the model's name.

▲ (Top) Porsche lowered windowsills 35mm, 1.4 inches, to make the sleek speedster. Base 60-horsepower "1600 Normal" Speedsters sold for 11,900DM ($2,833), while the company charged 13,000DM ($3,095) for the 75-horsepower 1600 Super Typ 616/2 engine.

▲ **1956 356A 1500 European Coupe** (Bottom) The 1,582cc "1600 Normal" developed 60 horsepower. This was good enough to accelerate the 850-kilogram (1,870-pound) coupes from 0 to 100 kilometers per hour in 16.5 seconds.

▶ Options either from the factory or well-stocked dealers included tinted sun visors and front wing windows. Base price of the "1600 Normal" was 12,700DM, $3,738.

Karmann discontinued its hardtop coupe for 1962 but joined Reutter in manufacturing standard coupes and convertibles. Porsche stopped producing the Roadster as well. Its hottest performer was the Carrera 2, using a 2.0-liter version of Fuhrmann's four-cam engine, offering customers 130 horsepower, in coupe or cabriolet form.

Reutter was a crucial partner to Porsche's present and future plans. However, when Ferry showed Reutter manufacturing projections for his new car, Reutter balked. The firm had lost its founder before the war and his son in a bombing raid. It would not accede to Porsche's needs in hiring additional help or financing new tooling. The family wanted out. Ferry faced an untimely choice, but he committed resources and acquired Reutter. The acquisition included costs of new tooling and a new building to house it and the manufacturing processes. As he neared his estimated production start, his investment approached DM 15 million, nearly $4 million at the time. For a small company, this represented an immense risk.

One of the largest obligations was to develop a new engine. Porsche customers admired the 130-horsepower, 2-liter flat-four Carrera 2 engines. But these were noisy and temperamental units—great for racing, with abundant power coming at high engine speeds, but ill suited to daily use on public roads. To achieve equal output from something quieter and more docile was a challenge Franz Reimspeiss handed off to another young engineer, Hans Mezger.

Reimspeiss first experimented with fuel injecting the 356 engine. This didn't develop the power Ferry demanded. Reimspeiss's next attempt, designated the Typ 745, was a 1,991cc opposed six with 80mm bore and 66mm stroke. Tests yielded 120 horsepower—close but no success. Enlarging bore to 88mm delivered 130 horsepower at 6,500 rpm from a 2,195cc package. But Reimspeiss and new engineering director Hans Tomala concluded that the 60mm stroke was too long for future racing development.

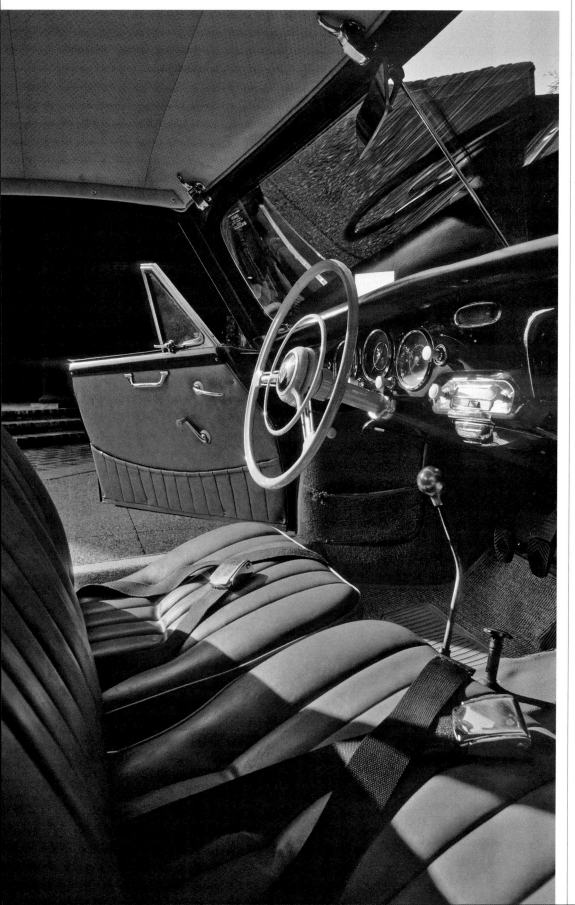

While engineers worked on the new flat six, the Racing Department moved ahead with ambitious plans for a flat eight designated the Typ 753. This complicated engine, with dual overhead cams, developed 185 horsepower (and more) from 1.5 liters (91.5 cubic inches). It required 220 hours for assembly, which made it far too costly for series production vehicles. Still, it yielded ideas.

Mezger commenced work on a Typ 821 engine with dual overhead cams and wet sump lubrication. Air-cooled engines reduced temperatures by circulating huge quantities of oil through them; however, the reservoir requirement for this engine proved too large to fit into F. A.'s car body. Mezger's next engine, the Typ 901, carried over many of the 821 and 753 developments, but it used dry sump lubrication pumped from a separate tank. This allowed Mezger to mount the engine lower in the car, improving handling and balance. It also made the new car's long, tapering roofline a certainty.

(CONTINUED ON PAGE 22)

◀ **1958 356A Hardtop-Cabriolet**
Three clamps above the windshield hold the front of the removable top onto the car body. The car weighed 885 kilograms (1,947 pounds) and the 1600 Normal engine accelerated it to 100 kilometers per hour in 16.5 seconds.

▶ (Top) Porsche introduced the removable hardtop cabriolet for 1958. Manufactured by Karmann of Osnabruck, the company charged 14,960DM ($3,041) for these truly "convertible" coupes.

▶ (Bottom) Porsche routed tailpipes through the bumper overriders on 1600 Normal models to improve rear ground clearance. Studs for the tonneau cover surrounded the removable top.

▲ One visible cue to identify the Carrera GT cars was the parallel hard oil lines inside the passenger front wheelwell. These potent 1,588cc (96.9-cubic-inch) Typ 692/2 models, using Ernst Fuhrmann's "four-cam" engine, developed 115 horsepower.

▶ **1959 356A 1600 GS Carrera GT Coupe**
Porsche sliced extra cooling louvers into the rear deck lid. With single outlet "stinger" exhaust, the company charged 18,500DM ($4,426). This exhaust system increased horsepower to 128 at 6,700 rpm.

◀ Countless racing careers were made in these interiors. Capable drivers could accelerate these coupes from 0 to 100 kilometers per hour in 11.0 seconds.

▲ 1960 356B 1600 Cabriolet
Known internally as the T-5 body, Porsche elevated the 356 headlights and front fenders as well as the front bumper. Air inlets below the bumper provided better front brake cooling.

(CONTINUED FROM PAGE 18)

In early 1963, another drivable prototype went to work, disguised with tail fins and painted in the olive drab of North Atlantic Treaty Organization (NATO) vehicles. It was nicknamed *der Fledermaus*, "the Bat," and used the Typ 821 overhead cam flat six. Five more prototypes followed quickly. Each was subtly different from the one before, with changes not only in engineering but in appearance, as F. A. Porsche and his modelers put finishing touches on the shape of the car.

Meanwhile, between June and Christmas 1963, F. A. Porsche developed a full line of new models, designated 901 for cars with a six-cylinder engine and 902 for those with a four-cylinder. He created a sunroof coupe, a second

▲ (Top) The B series introduced a new steering wheel with the option of this three-spoke wheel from Nardi. New front seats offered more comfort, and Porsche shortened the gearshift lever 40mm (1.6 inches).

▲ (Bottom) F. A. Porsche started developing the 901/911 from this four-seater proposal that the engineering department created in 1952. Its 2,400mm (94.5-inch) wheelbase was 300mm (11.8) inches longer than the 356 and 189mm (7.4 inches) longer than first-generation 901/911 models.

model with a removable roof panel, and a third as a full cabriolet. His Cabrio concept, however, left insufficient space for the cloth roof to fold on top of the 2-liter flat six.

One of F. A. Porsche's cousins was Ferdinand Karl Piëch, the son of Ferry's sister Louise. He graduated in 1962 from Eidgenössische Technische Hochschule in Zurich, with a master's degree in mechanical engineering, and he joined his uncle's company immediately. He supervised 901 engine development. It was a vexing process. An assortment of engines and drivetrains went into and came out of the seven 1963 prototypes, and two more appeared in April 1964. It wasn't until early 1964 that dry sump 901 flat sixes were ready for road work. By this time, two development 902 models had appeared as well, running a variety of four-cylinder engines.

For the flat sixes, Mezger and Piëch selected three-barrel Solex overflow carburetors over each bank of cylinders. These provided one float chamber for all three 30mm venturis. This system managed the flow of fuel from the gas tank, rather than what flowed through the jets into the cylinders. It eliminated a risk for enthusiastic Porsche drivers of floats that hung up in centrifugal force while cornering hard. The public already had seen the car, prototype 5 (without a working engine), assembled at Karmann and painted yellow, on Porsche's show stand at the Frankfurt IAA show starting on September 12, 1963. Order books filled with deposits from those willing to wait at least a year for delivery.

To keep interest in Porsche's cars alive (and much-needed revenues coming in), the company introduced its ultimate 356, the C series, in late 1963 as 1964 models. Base versions included a 75-horsepower flat four, while the

CONTINUED ON PAGE 27)

▲ (Top) The company offered Super 90s as coupes, cabriolets, and the Karmann-built hardtop coupe and hardtop cabriolet configurations. These cabriolets weighed 900 kilograms (1,980 pounds).

▲ (Bottom) Porsche quoted a top speed of 180 kilometers (112 miles) per hour. The S90 cabriolet sold for 16,950DM, $4,216.

▶ **1961 356B S90 Roadster**
Porsche used its 1,582cc (96.5-cubic-inch) Typ 616/7 flat four to power the Super 90, so named for its horsepower output at 5,500 rpm. Acceleration from 0 to 100 kilometers per hour took 13.5 seconds.

▲ **1962 356B Carrera 2 Cabriolet**
For several years this car was Ferry Porsche's daily driver. It was a test bed for numerous engineering and design updates, yet, like its driver, it was understated.

◄ According to the history of this car, F. A. Porsche used his father's car as a development prototype, fitting it with the T-6 bodywork that appeared in final 356B models and all 356C versions.

▶ **1962 754T7 Wind Tunnel Model**
The origins are clear, from the headlight and front fender treatment to the sloping roofline and the angle to the taillights. F. A. Porsche created the forms with modelers Heinrich Klie and Gerhard Schröder starting in 1959.

(CONTINUED FROM PAGE 23)
SC or Super C elevated the mid-range 90-horsepower engine of the B series to 95 horsepower. The Carrera 2 still provided 130 horsepower in a coupe or cabriolet. Porsche's four-wheel ATE disc brakes were a significant equipment upgrade for aggressive drivers who had always appreciated the company's big drum brakes.

Porsche continued to manufacture 356C and SC models through the 1965 model year and assembled nearly 16,700 coupes and cabriolets. Manufacture ended in early summer; however, Porsche produced 10 more C cabriolets for the Dutch national police early in 1966.

By then it was selling its 901 and 902 models, having begun in September 1964. (Helmuth Bott's engineers had completed two final "preproduction" cars—the 12th and 13th prototypes, given production serial numbers 300001 and 300002—for durability testing in mid-September.) Ferry Porsche had not only set the stage for his 356 successor, he had raised the curtain. The first customer car, 300007, rolled out of the Zuffenhausen plant on September 14, 1964. The outside world responded; rave reviews reached Zuffenhausen.

▲ (Bottom) **1962 754T7 Wind Tunnel Model**
The T-7 went through several iterations as a 2+2 and a four-seater. Beside the taillights were vents to reduce engine heat.

▲ This experimental 1,966cc (119.9-cubic-inch) flat four developed 130 horsepower at 6200 rpm. Ferry Porsche ran this in his personal car for several years.

▶ **1964 356C 1600 Coupe**
Porsche carried over the T-6 body it had introduced on 356B models starting with model year 1962. The 1,582cc Typ 616/15 engine developed 75 horsepower at 5,200 rpm.

▲ **1964 356C 1600 Coupe**
The 356C accelerated from 0 to 100 kilometers per hour in 14 seconds and reached a top speed of 175 kilometers (109 miles) per hour. The car sold for 14,950DM, $3,756.

▲ **1965 356SC 1600 Cabriolet**
Porsche's sportier SC with the Typ 616/16 engine offered buyers 95 horsepower at 5,800 rpm, enough to get the cars to 100 kilometers in 11.5 seconds. The cabriolets sold for 15,950DM, $3,997. Coupes went for 14,950DM, $3,883.

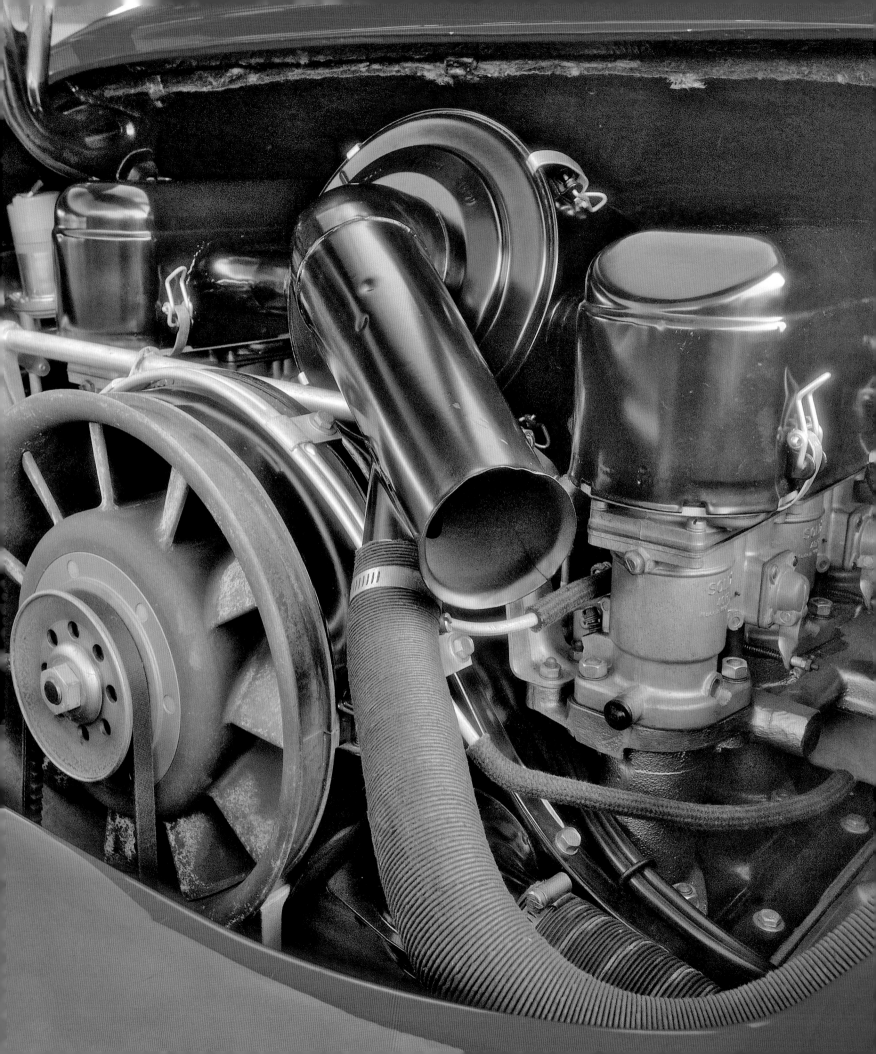

1964–1965 **901**
1964–1967 **911**
1967–1968 **911S**
1967 **911R**
1968 **911L**
1965–1969 **912**

CHAPTER 1

THE FIRST GENERATION 1964–1969

◀ 1964 901 Production Coupe
The snorkel air intake arrived with production versions of the 901. This engine was designated the Typ 901/01 and with the clutch it weighed 184 kilograms (405 pounds.)

1964-1965 901

Porsche's 901 and 902 seized the motoring world's attention. As Ferry and his marketing staff showed the prototype to visitors at the Frankfurt show, his engineers readied the car for manufacture. When the exhibition closed, sales and marketing staffs celebrated full order books. They returned to Stuttgart to prepare for the Paris Auto Salon in October. In France crowds again thronged Porsche's booth, and sales representatives recorded names and deposits. Not everyone was thrilled with the new models, however. After the salon ended, Automobiles Peugeot notified Porsche that it could not use the 901 and 902 designations in France.

Despite producing vastly different automobiles, Peugeot believed it had cause for complaint. Back in 1929, the French carmaker had introduced its Model 201, a low-priced vehicle intended to expand sales to a broader range of customers. Soon afterward the company registered with the French copyright and patent office its right to use three-digit model designations that incorporated a 0 in the middle. The first car, the 201, was a 2 series chassis with a 1.1-liter engine. Peugeot progressed through 301, 401, and 601 series, arriving at the 403 and 404 when Porsche showed the 901 and 902.

That Porsche manufactured essentially two-seat sports cars and Peugeot produced utilitarian sedans, coupes, and station wagons (and that Porsche already had raced its Carrera 904GTS at Le Mans and had won the French Grand Prix at Rheims in its 804) were matters

▲ **1964 901 Prototype**
Two young engineers, Hans Mezger and Ferdinand Piëch, developed the dual overhead camshaft engine to produce 130 horsepower at 6,100 rpm. The Typ 901/01 engine displaced 1,991cc (121.5 cubic inches) and used two three-barrel Solex 40 PI carburetors.

▶ Porsche assembled 13 prototype 901 models from 1962 through 1964. This car, number 7, is the only one known to exist. Its center crank sunroof existed only through prototypes.

Ferry Porsche decided to ignore. France represented a large market for his cars, so he changed the designation. It had come about not from any conscientious pursuit of a distinctive number but out of convenience. Porsche had entered a parts distribution partnership with Volkswagen, which had developed a complex numbering system across its model lines. The sequence available for Porsche started with 901. Internally, the company kept its parts nomenclature, but according to historian Tobias Aichele, to provide the motoring world a consistent name for its models, Porsche switched to 911 and 912 on November 10, 1964, Porsche had manufactured 82 901s by this time, scheduled for customer delivery at the end of the month. The company rebadged all of them.

▲ For the 901, the instrument panel was still a work in progress. Its roots from the 356 are clear, from two big dials to ignition key placement.

▲ Ferry Porsche first asked Erwin Komenda to make prominent front fenders for the 356 so he could tell where the front wheels were. That characteristic directed creators of this 901 as well.

▶ Dual rear exhausts hinted at performance options to come. This 901 prototype differs in many small ways from 911 production beginning in August 1964.

THE FIRST GENERATION 1964–1969 | 37

▲ **1964 901 Production Coupe**
Porsche began manufacturing the 901 in August 1964 and had finished assembly of 82 cars before Peugeot complained about the designation number. On November 10, 1964, the company invented the 911 designation, but most people refer to those first cars as the 901.

◀ The Petersen Museum in Los Angeles, California, owns this painstakingly researched and carefully restored example. The 901 sold for 21,900DM ($5,503) at the factory.

▲ Designer Otto Soeding, who created the curved five-instrument panel, explained that Erwin Komenda had him mimic the curve of a large ball. That way all gauges faced the driver and glare would obscure only one at a time.

▶ **1965 911 Cabriolet Prototype**
Shortly after production began on the 901, Ferry Porsche ordered another cabriolet prototype prepared. The question they faced was where to put a retractable top if they proceeded with manufacture.

▼ While the decision to use a dry sump engine allowed engineers to mount the flat six lower on the platform, there still was not enough space for a cavity to store a collapsible top. Redesign and tooling of new rear quarter panels would have cost a fortune.

▲ Engineering developed an experimental windshield header to attach a retractable top. Slicing off the metal roof proved the shortcomings of the original design that left room only to set the collapsed cloth roof on top of the bodywork.

▲ The owner of this car delighted in driving into Porsche gatherings and standing at a distance listening to comments. While it appeared someone merely had sliced off a Targa bar, few guessed it was done in late 1964.

YEAR	1964-1965
DESIGNATION	901
SPECIFICATIONS	
MODEL AVAILABILITY	Cabriolet prototype
WHEELBASE	2211mm/87.0 inches
LENGTH	4163mm/163.9 inches
WIDTH	1610mm/63.4 inches
HEIGHT	1320mm/52.0 inches
WEIGHT	unknown
BASE PRICE	unknown
TRACK FRONT	1337mm/52.6 inches
TRACK REAR	1317mm/51.9 inches
WHEELS FRONT	4.5Jx15
WHEELS REAR	4.5Jx15
TIRES FRONT	165-15 radial
TIRES REAR	165-15 radial
CONSTRUCTION	Unitized welded steel
SUSPENSION FRONT	Independent, wishbones, MacPherson struts, longitudinal torsion bars, hydraulic double-action shock absorbers
SUSPENSION REAR	Independent, semi-trailing arms, transverse torsion bars, hydraulic double-action shock absorbers
BRAKES	Discs, 2-piston cast-iron fixed calipers
ENGINE TYPE	Horizontally opposed DOHC six cylinder Typ 901/01
ENGINE DISPLACEMENT	1991cc/121.5CID
BORE AND STROKE	80x66mm/3.15x2.60 inches
HORSEPOWER	130@6100rpm
TORQUE	128lb-ft@4200rpm
COMPRESSION	9.0:1
FUEL DELIVERY	2 Solex 40 P-I overflow carburetors
FINAL DRIVE AXLE RATIO	4.428:1
TOP SPEED	130mph
PRODUCTION	1 Typ 901 prototype cabriolet 82 Typ 901 production coupes

1964-1967 911

Then, following a polite span of time, Baron Huschke von Hanstein, Porsche's racing and public relations director, wrote the Fédération Internationale des Automobiles (FIA), the world's auto racing governing body, based in Paris. He noted that Porsche's racing models, not available to regular customers, probably did not infringe on Peugeot's right to use the middle 0, thus enabling Porsche to retain the 804 and 904 model designations. The FIA agreed, whereupon von Hanstein announced development of a new racing model known as the Typ 906 Carrera 6.

While engineers put finishing touches on the 911 prototypes, the company completed assembly of the 904s it sold to qualified racers and of the last 356-1600C and SC models introduced at Frankfurt alongside the 901. Reutter, Karmann, and Weinsberg body works hustled to meet Porsche's production needs. As production began, Karmann assumed a larger role, assembling and trimming completed 911 car bodies. When early magazine reviews praising the new model poured in, Ferry discontinued 356C and SC manufacture at the end of the 1965 model year.

Ferdinand Piëch and engine designer Hans Mezger initially conceived a four-cylinder engine based on the Typ 745 2-liter, six-cylinder 911 engine for use in the third line. This was the entry level or standard version that started out as the 902 but became the 912 following Peugeot's request. The concept was commonality of parts, but Porsche had thousands of four-cylinder, 1,582cc Typ 616 engines left. These engines had 82.5mm bore and 74mm stroke; in the SC, they developed 102 DIN horsepower. Porsche planned to use them in industrial power units, but when that project collapsed, the company halted Typ 745 four-cylinder development and installed the earlier four in the 902/912, but with 9.3:1 compression, good for 90 DIN horsepower at 5,800 rpm and 90 lb-ft torque at 4,200 rpm. Sales director Wolfgang Raether and sales manager for Germany Harald Wagner trimmed luxuries on the 912 to hold its price near that of 356C coupes.

By the end of the 1965 model year, Porsche had produced 3,390 911 models, but it had manufactured 6,401 912s. Customers enthralled with the 911s looks at the Frankfurt and Paris shows saw they could have the appearance and handling for nearly $1,400 less if they accepted a lower-powered engine. (They also got less weight; without the dry sump and with two fewer cylinders, 912s came in at 995 kilograms/2,190 pounds, which was 100 kilograms—220 pounds—lighter than the new 911.) Most early 912 purchasers were European; to accommodate loyal customers on the continent, Porsche delayed exporting 912s to North America until it ended 356C production in September 1965. Americans got the 912 at the start of the 1966 model year.

Ferry Porsche's first car was a roadster. Soon after he introduced Gmünd coupes, he offered open cars without interruption, until the 911 arrived. Ferry and Wolfgang Raether conceived the 911 as a successor to the 356 with a full range of models. F. A. Porsche's design staff, particularly Heinrich Klie and Gerhard Schröder, created models and drawings supplementing the coupe in late 1963. The principal problem was that the car had no room to store the collapsible top. Engineer Helmuth Bott's initial efforts to remedy the new car's handling challenges made clear that removing the roof was not sensible. That large steel panel provided much of the body stiffness that helped Bott's improvements work effectively. The 911 tub simply was not rigid enough to support a cabriolet body.

Ferry did not give up. He planned a sunroof version, one with a removable roof, a cabriolet, and a deluxe model. This was to debut at Frankfurt, fitted with a leather interior. He also envisioned a Sport or S model with 150

▲ The wood-rimmed steering wheel and dashboard were standard equipment. The five-speed transmission placed first gear to the left and back.

horsepower to top out the line. As development advanced on the coupes, Ferry Porsche approached Reutter Carrosserie's Walter Beierbach, asking for a cabriolet body based on the Typ 745 T7 styling proposal that F. A. Porsche's styling staff had done. This successful prototype handled well but still lacked space for a collapsible top.

A year later, drawings and a model went to Karmann in Osnabruck. Could they evaluate what it would take to manufacture an open 911? F. A.'s first variation incorporated suggestions from staff designer Gerhard Schröder on convertible folding systems and collapsible bows. The car appeared with a folding padded top that was stored below a boot. It left a startlingly low silhouette, but it reduced the size of the engine compartment. The second version provided owners a removable cloth top to stretch across bows as customers had done with 356 Speedsters. This kit came out of a smaller boot storage area, but still it intruded into the engine compartment. The third option was a detachable two-piece roof comprising one panel over the driver and passenger and a second for the rear window. F. A. Porsche envisioned these attached to a rigid rollover bar or, alternatively, one that collapsed, so that the car presented a clean profile.

F. A. Porsche preferred the true cabriolet. As he explained in an interview in 1991, "Open cars at Porsche always have been roadsters or speedsters. These have followed a distinct shape for the roof that is not the same as our coupes. I wanted a clear break in the roofline to reemphasize that roadster character." His revised form did not duplicate the elegant sweep of the coupe. It would have required expensive new body panels for the rear of the car. For a car already costing his father millions, these new expenses killed the cabriolet.

▲ **1965 911 Coupe**
Porsche's first production 911s reached U.S. buyers in February 1965. The company charged American customers $6,500 for the car.

◀ The production 911 weighed 1,080 kilograms, 2,376 pounds, and at the factory it sold for 21,900DM, $5,503. The company manufactured something like 3,389 of these cars.

▲ 1966 Bertone 911 Roadster
Its elegant lines may have been a bit too Italian for Porsche. The company also worried about assembly quality since the car was called a Porsche. Von Neumann commissioned only one prototype.

Development and production charges tripped up the concept of a collapsible roof bow/rollover bar as well. It required chassis and body modifications that F. A. Porsche had not imagined. He was a designer, not a production body engineer. Erwin Komenda, who was the company's body engineer, never warned him about these needs when they collaborated on engineering drawings for 911 bodies. In the end, there was no room to store the roof structure without making expensive changes in design, engineering, and tooling. Gerhard Schröder understood where problems were, based on experience with 356 series convertibles and roadsters. As he explained to former Porsche press staffer Tobias Aichele, "With the engine in place, the folded top stood above the body like a VW cabriolet and that would not have been acceptable for a sports car of Porsche's stature." The 911 shape left only one choice at this point: A center bow welded rigidly in place would support F. A. Porsche's third concept, a top with two removable panels. This bow earned engineering support because it increased stiffness to the open car body.

Helmuth Bott, in an interview decades later, recalled another factor influencing the open car. "We were worried that the American Congress might legislate against convertibles. That had been a big portion of our market for open cars." Consumer safety activist Ralph Nader took General Motors and Chevrolet Division to task for pinching pennies, thereby compromising the handling of Chevy's rear-engine air-cooled Corvair. GM engineers

▲ California Porsche distributor Johnny von Neumann had great success with Max Hoffman enticing Porsche to produce an open car for American enthusiasts. The next time von Neumann got more ambitious and developed this Italian roadster design from Bertone by himself.

argued for—but accountants denied them—anti-sway bars to control the rear independent swing axles, similar to what Porsche and Mercedes-Benz had used for years without incident. Questions about Corvair safety awoke lawmakers, instigated hearings, and threatened to legislate the kinds of cars Americans could drive.

Porsche completed a styling prototype in mid-June 1964. In late January 1965, Bott drove an engineering model with additional reinforcement at the base of the A-pillar. "It was," he recalled, "as stiff as I expected, about as the current three-five-six. . . . In time it became much better."

Porsche's patent expert, Emil Soukup, knew of an innovation that England's Triumph had introduced on its TR4 sports car in 1962. Italian stylist Giovanni Michelotti had created a hardtop from which owners could remove a solid center section. Porsche's concept was different enough that Soukup registered their design. During a meeting with his staff, as they discussed how automakers appropriated racetrack names for car models, sales director Harald Wagner came up with the name "Targa." Porsche had used "Carrera" to honor its accomplishments in the Mexican road race. "Targa" was a recognizable word and easily pronounced in all languages. Its translation, "shield," fit with F. A. Porsche's plan to emphasize the central bar's function by finishing it in brushed stainless steel. The company began manufacturing the Targa as a 1967 model, allocating 15 percent of production, 7 out of 55 cars a day, to the new body style.

The increasing popularity of the 911 and 912 ensured their success and allayed Ferry Porsche's lingering fears. Production set a new record at 12,820 cars. Of these, 9,090 were the affordable 912s. The company shipped nearly half of its production to the United States. Porsche had completed its own emissions test facility, the first one in Europe that the U.S. Environmental Protection Agency (EPA) approved for vehicle certification over its official driving cycles.

It is a part of 911 history that innovators, entrepreneurs, and outsiders believed they could build a better car. With the new model, one man wasted little time. Johnny von Neumann owned a business in Los Angeles, selling Porsches he got through U.S. distributor (and fellow Austrian) Max Hoffman. As early as 1953, von Neumann gave Hoffman ideas for and opinions on Porsche models. Back then, von Neumann believed his customers wanted "a boulevard racecar" to cruise Sunset Strip, with an arm on the doorsill while they gawked at girls. In September 1954, Porsche introduced the Speedster, and initial production came to Hoffman. Johnny shuttled them across the country to his customers. They sold to racers and movie stars, becoming an automotive icon and giving von Neumann credibility, even as Hoffman took credit for the idea. The factory denied Speedster access to Europeans for the first nine months.

By October 1965, von Neumann had seen the 911 and knew about the Targa. It wasn't enough for him, but this time, unwisely, he bypassed Hoffman and went directly to Turin, Italy. He contracted with designer Nuccio Bertone to create a new roadster on the 911 platform, with a view toward series production of 100 cars or more. Then von Neumann went to Porsche to arrange for a quantity of platforms with engines and full wiring.

The idea met resistance. Ferry Porsche and Harald Wagner had calculated that the car would have to sell for $7,000 or more. While that seemed costly, Ferry had other concerns. "It has our name on it," he told von Neumann, "so we are concerned to be sure it is going to be right." Von Neumann recalled that, at that moment, he felt the project could not succeed, no matter how it might look.

Bertone showed the finished prototype on its stand at Geneva in March 1966. It had finished the car just hours before the show doors opened. On its stand, Porsche displayed the Targa with its body-strengthening roll bar.

"The thing that killed it," von Neumann explained 35 years later, "was something they already knew. When you make a convertible out of a coupe, there's some chassis movement, flexing." During the show and in the months after, Bertone and von Neumann fielded inquiries. But no orders materialized. Where Count Albrecht Goertz's 695 prototype was "too American," the Bertone roadster may have been too Italian. Or the audience at Geneva saw the Targa as Porsche's open 911.

◀ **(Opposite) 1966 911 Coupe**
After experiencing trouble with soft throttle bodies on the Solex carburetors, in February 1966, Porsche switched to Weber 40 IDA models. Engine designation changed to 901/05.

▼ Porsche quoted acceleration from 0 to 100 kilometers per hour at 9.1 seconds from the 2-liter flat six. Top speed was 210 kilometers (130 miles) per hour.

▲ The Typ 901/01 engine had 80mm (3.15 inch) bore and 66mm (2.6 inch) stroke. Nine liters of oil cooled and lubricated it.

▲ With few visible changes to the car, Porsche increased the price for 1966 models by 1,000DM ($250) to 22,900DM ($5,725). The company manufactured just 1,709, though it introduced the popular 912 the same year.

YEAR	1964–1967
DESIGNATION	911
SPECIFICATIONS	
MODEL AVAILABILITY	Coupe; Targa introduced in 1967
WHEELBASE	2211mm/87.0 inches
LENGTH	4163mm/163.9 inches
WIDTH	1610mm/63.4 inches
HEIGHT	1320mm/52.0 inches
WEIGHT	1080kg/2376 pounds
BASE PRICE	$5,489
TRACK FRONT	1337mm/52.6 inches
TRACK REAR	1317mm/51.9 inches
WHEELS FRONT	4.5Jx15
WHEELS REAR	4.5Jx15
TIRES FRONT	165-15 radial
TIRES REAR	165-15 radial
CONSTRUCTION	Unitized welded steel
SUSPENSION FRONT	Independent, wishbones, MacPherson struts, longitudinal torsion bars, hydraulic double-action shock absorbers
SUSPENSION REAR	Independent, semi-trailing arms, transverse torsion bars, hydraulic double-action shock absorbers
BRAKES	Discs, 2-piston cast iron fixed calipers
ENGINE TYPE	Horizontally opposed DOHC six cylinder Typ 901/01
ENGINE DISPLACEMENT	1991cc/121.5CID
BORE AND STROKE	80x66mm/3.15x2.60 inches
HORSEPOWER	130@6100rpm
TORQUE	128lb-ft@4200rpm
COMPRESSION	9.0:1
FUEL DELIVERY	2 Solex 40 P-I overflow carburetors
FINAL DRIVE AXLE RATIO	4.428:1
TOP SPEED	130mph

▲ 1967 911S Coupe
Bruce Jennings was one of America's most proficient 356 and 911 racers. Hungry for a special car, he special ordered this S, painted in Cadillac Bronze Metallic. To fit oversized 5.50 Goodyear Bluestreak tires, he "rolled" out the fender lips.

1967-1968 911S

Through all this, Helmuth Bott divided his time between making a stronger, safer open 911 and developing one that met the expectations of Porsche's performance enthusiasts. The new sport model S was more luxurious, but most of its upgrades were inside the engine. Engineer Paul Hensler enlarged valve diameters and extended valve timing to allow more fuel into each cylinder and extract exhaust more effectively. His work boosted engine output to 160 brake horsepower (DIN), 180 SAE gross. Its fan operated inside a distinctive red shroud.

Weber carburetors helped accomplish the horsepower increase. Porsche determined that throttle bodies of the Solexes were too soft to hold tune, so engineers specified the Italian carburetors for all 911 models. For the new S, they selected 40IDS3C models with 32mm venturis. Base models used 30mm-venturi versions of the carburetors.

Porsche introduced Fuchs forged aluminum wheels, which circulated more air to cool the brake rotors. The aluminum saved nearly 5 pounds (2.3 kilograms) per wheel over the previous steel ones. Bott's engineers installed larger front anti-sway bars and added rear ones, as well as adjustable Koni shock absorbers. F. A. Porsche's designers gave the interiors full leather or vinyl seats, with corduroy or houndstooth cloth inserts. Porsche discontinued wood trim instrument panels and wrapped the steering wheel in leather.

At Max Hoffman's dealership in New York City, the base 911 sold for $5,990 with a four-speed transmission. The four-cylinder 912 went for $4,790. The S was $6,990. The Targa added $400 to the price of any model, Fuchs wheels were another $375, and an optional five-speed transmission cost $80. Factory prices ranged from DM 16,980 for the 912 to DM 20,980 for the base 911 and DM 24,480 for the 911S.

The S was Porsche's hottest 911 model, racing from 0 to 60 miles per hour in 6.5 seconds, according to tests by *Car and Driver* magazine, incredibly quick for 1967 and comparing favorably with America's muscle cars and Corvettes. The S topped out at 140 miles (225 kilometers) per hour, 10 faster than the base model. But European and U.S. magazines found the autobahn-oriented S models less satisfying to drive in urban areas, where high-torque and -horsepower curves didn't fit comfortably with stop-and-go driving.

▲ Porsche responded to requests for an open 911 with a Targa that incorporated removable panels for the rear window and over the driver's head. The 911S introduced Fuchs wheels to Porsche owners. Targa sold for 25,880DM, $7,390 in the United States.

▶ Over 35 years of racing, Jennings won 216 events and finished second another 104 times. He not only specified the exterior color but also had the leather-covered Recaro sport seats upholstered to match a sample he sent.

Production of the 911 reached only 1,709 cars for 1966, while assembly totals surged for model year 1967 with the introduction of the Targa and S models. Output was 11,011 cars; 6,472 of them were 912s, 1,823 were S Coupes, and 1,201 were Targas in both base and S versions. Once again, nearly half—about 5,400 cars—went to U.S. dealers and customers.

A Rally Kit, option 9552, replaced standard seats with a pair of Recaro seats and installed a roll bar, a 100-liter fuel tank with through-hood fill, adjustable Koni shock absorbers, and a litany of engine modifications and weight-loss provisions. These cars came with no rear seat, no carpets, and a smaller steering wheel, as well as an encyclopedia of other options and standard-equipment deletions.

Sport Kit I gave the S engine larger main jets and carburetor chokes, lower restriction air cleaners, and hotter spark plugs, all of which boosted output to 170 DIN horsepower from 160. Sport Kit II replaced the exhaust system and many glass and steel panels with lighter materials. Porsche encouraged its American dealers to promote these options, and racers up and down both coasts and across the middle of the country became steady buyers of cars, options, and upgrades, and took home the trophies to prove it.

The model line expanded again for 1968 as Porsche supplemented the 912, 911, and 911S with a new 911T series. The company conceived of this model as a six-cylinder version it could sell in Germany for less than DM 20,000 (about US$5,000 at the time). Engineers

▲ Jennings specified simple rain shields over the carburetors, replaced later with low-restriction air cleaners over his 46mm Weber 3/3 C1 carburetors. Stock Typ 901/02 engines developed 160 horsepower at 6,600 rpm, providing acceleration from 0 to 100 kilometers per hour in 7.6 seconds.

fitted the car with a four-speed transmission and solid disc brakes, and the company deleted many of the luxury and trim features it provided on the base 911. The car weighed about 77 pounds (35 kilograms) less than the base model. This provided the Competition Department a new platform on which to develop racing and rally models. Its standard engine (with a black fan shroud) developed 110 brake horsepower DIN (125 SAE), and German magazines got 0-to-60-mile-per-hour times of 8.1 seconds, with top speeds of 129 miles per hour (206 kilometers per hour).

▲ **1967 911S Targa**
The company sold the S Targa for 25,880DM (roughly $6,486,) and $7,390 in the United States. The Weber-carbureted Typ 901/02 engine developed 160 horsepower at 6,600 rpm.

YEAR	**1967-1968**
DESIGNATION	**911S**
SPECIFICATIONS	
MODEL AVAILABILITY	Coupe, Targa
WHEELBASE	2211mm/87.0 inches
LENGTH	4163mm/163.9 inches
WIDTH	1610mm/63.4 inches
HEIGHT	1320mm/52.0 inches
WEIGHT	1030kg/2266 pounds coupe – 1080kg/2376 pounds Targa
BASE PRICE	$6,153 coupe – $6,486 Targa
TRACK FRONT	1337mm/52.6 inches
TRACK REAR	1317mm/51.9 inches
WHEELS FRONT	4.5Jx15
WHEELS REAR	4.5Jx15
TIRES FRONT	165-15 radial
TIRES REAR	165-15 radial
CONSTRUCTION	Unitized welded steel
SUSPENSION FRONT	Independent, wishbones, MacPherson struts, longitudinal torsion bars, hydraulic double-action shock absorbers
SUSPENSION REAR	Independent, semi-trailing arms, transverse torsion bars, hydraulic double-action shock absorbers
BRAKES	Discs, 2-piston cast iron fixed calipers
ENGINE TYPE	Horizontally opposed DOHC six-cylinder Typ 901/08
ENGINE DISPLACEMENT	1991cc/121.5CID
BORE AND STROKE	80x66mm/3.15x2.60 inches
HORSEPOWER	160@6600rpm
TORQUE	132lb-ft@5200rpm
COMPRESSION	9.8:1
FUEL DELIVERY	2 Weber 40 IDS carburetors
FINAL DRIVE AXLE RATIO	4.428:1
TOP SPEED	140mph
PRODUCTION	3,573 coupes; 925 Targas; 1 4-door sedan

1967 911R

The lighter platform of the 911T, developed in 1967, led to many competition variations. These cars mostly remained in Germany and Europe, special creations from Ferdinand Piëch's Experimental Department, Peter Falk's Competition Department, and Hans Mezger's racing engine shop. Few outsiders recognized these designations until years later. These cars, the 911R in 1967 and the 911ST, 911GT, and 911T/R in 1968 and later years, were meant for Europe's endurance races, speed record trials, and long-distance rallies. Their successes helped develop the mystique of the 911.

The 911R, the *rennsport* or "race" model, resulted from Ferdinand Piëch's efforts to develop a lean 911. His engineers assembled four prototypes, each successive car using thinner fiberglass panels and Plexiglas windows, as well as metalwork perforated to remove weight. Racing mechanic Rolf Wütherich developed a device that precisely placed and drilled holes through almost any body panel, seat rail, or foot pedal. He patiently shaved away ounces from each car.

Perhaps most significantly, the first "production" R delivered an unexpected triumph to Porsche. High-speed-record endurance runs were obscure challenges that manufacturers forced on each other. Ford's GT40 held a record for a while, running 112.5 miles per hour (180 kilometers per hour) over four days around Daytona in December 1965, totaling 12,500 miles (20,000 kilometers) at the end. Then Toyota averaged 128.75 miles per hour (206 kilometers per hour) around the Fujiyama circuit, setting 6,250- and 9,375-mile (10,000- and 15,000-kilometer) records. Four Swiss racers calculated that using the Carrera 6 long-tail racer owned by two of them, they could beat all these performances. They applied to the FIA and booked time at Italy's Monza circuit. Monza had banked walls at either end of an infield oval, but its weather-damaged concrete surface broke the 906's upper suspension mounts within hours of their start. Rules required entrants to carry all spare parts and necessary tools in the car during the run. Only oil, gas, spark plugs, spare tires, and jacks could remain in the pits. The added weight proved to be too much load, and after just 625 miles (1,000 kilometers), co-driver Jo Siffert pitted with a broken shock absorber mount.

It's possible the R series was just Ferdinand Piëch's experimental test fleet and that he didn't care whether Porsche built or sold more of them. In any event, Wagner and Raether had

▼ 1967 911R Coupe
Front and rear deck lids were thin fiberglass. Side windows were 2mm Plexiglas and the windshield was only 4mm thick. Under the hood was a 100-liter (26.4-gallon) fuel tank.

seen the potential accurately; some of the original 1967 production 911Rs still were available as late as 1970.

Porsche's 1967 Sportomatic transmission was a sales success. The company had witnessed the arrival at driving age of an entire U.S. generation that could not manually change a gear or operate a foot clutch. A decade earlier, enthusiasts believed an automatic transmission did not belong in a sports car, though some American manufacturers tried. But by the late 1960s, Porsche had a new concept and a new audience. In the Sportomatic semiautomatic transmission, engineers mated a four-speed manual gearbox to a hydraulic torque converter through a clutch that disengaged automatically the instant the driver touched the gear lever. It was not a true automatic in that the driver had to change gears or risk over-revving the engine. Ironically, Porsche's European customers were first and most enthusiastic in adopting the new transmission. For years Porsche delivered many more Sportomatics to Europeans than to Americans.

It took an American imagination to develop a unique 911S and to take it beyond prototype stage. A Texan from San Antonio, William Dick, was a VW and Porsche distributor for Texas, Oklahoma, New Mexico, Wyoming, and Colorado. While he drove Ferraris and Maseratis, he wanted something more practical. He sent his general manager to Europe to interview independent designers willing to take on his idea—he wanted a four-door 911S for his wife, Hester. He had to go to Los Angeles to find a shop willing to follow through.

▼ Engineer Ferdinand Piëch was obsessed with vehicle weight. His staff pared the 911R down to 800 kilograms, 1,764 pounds. This was 230 kilograms, 506 pounds, less than the production 911S coupe.

▲ Following a development run of four prototypes, Porsche contractor Karl Baur in Stuttgart assembled 21 "production" versions of the lightweight 911R. This car was most notable for a number of endurance records it set under BP sponsorship.

▲ While rear doors offered easy access, seating was confined because the low 911 seats provided no room for toes beneath them. Rumors suggested the conversion cost was more than $20,000 in 1968.

Coachbuilder Dick Troutman and mechanical engineer Tom Barnes in Culver City, California, had a history of designing and producing original race cars for very successful independent American teams. They agreed to build Dick's car. In December 1966, an early production S arrived in their shops. Metalsmith Emil Diedt removed the roof and lengthened the platform 21 inches (53.3 centimeters) just in front of the B-pillar. The result was a 108-inch (2,743mm) wheelbase. Stylist Chuck Pelly sketched lines of the new car on plywood and sheet metal, as if it were a prototype destined for its racing debut a month later. He redesigned the front fenders to lay the headlights farther back and made other changes, both minor and bold. As Troutman and Barnes neared completion in August 1967, they adopted two new Porsche options, Fuchs wheels and the new Sportomatic transmission. Dick asked the builders to paint the car dark green.

According to Porsche historian Wolfgang Blaube, the four-door found its way onto the Supervisory Board agenda in Stuttgart in the summer of 1968. In considering the viability of a four-door model, the company had hired Italian designer Sergio Farina at Carrozzeria Pininfarina to come up with an idea, code-named 911/B17. It looked swollen and bilious in its psychedelic green paint, and in November 1969 the board passed on the idea. (Ferry then asked his styling department to design another one, coded 911/C20 and given the sequence number Typ 915. Stretched by 13.5 inches, [34.3 centimeters] and completed in July 1970, it was somewhat less appealing than the Pininfarina version. By that time, new man-

▲ To accommodate four doors, T&B purchased two additional 911 doors and mounted them backwards as "suicide doors" to provide rear seat access. The B-pillar was robust; the car did not creak with all four doors open.

▲ Texan William Dick contracted race car builders Troutman and Barnes in Culver City, California, to make a four-door 911 for his wife. T&B lengthened the wheelbase 533mm, 21 inches, to provide room for rear seats and foot wells.

▲ **1967 911 Four Door**
Troutman and Barnes built the car with the Sportomatic Transmission, a newly arrived option. In later years, the owner replaced it with the S five-speed.

▲ Four drivers shared responsibilities to get the lightweight car around the Monza banked circuit. Lapping the circuit in 1 minute 11 seconds, they averaged 209 kilometers per hour. Each driver took a two-hour shift. *Photograph courtesy Porsche Archive*

agement was in place, and another idea, for a larger car with a front engine and water cooling, arose. Ernst Fuhrmann gave that the number 928.)

One other 1967 project had even greater long-term impact than the Dick four-door, the 911R, or the Sportomatic transmission. Back in late 1960, Porsche had acquired a 93-acre parcel of rolling farmland near the village of Weissach. Helmuth Bott designed a 1.8-mile test track and skid pad for the property. In 1967 Ferdinand Piëch launched a $20 million building program, intending to put "everyone needed for design and engineering of a car within one hundred meters of each other." (Since then, the campus has stretched that distance to 300 meters.)

The company's research and development engineers and designers had plenty of outside work to support the facility. Volkswagen used Porsche for its R&D, a carryover from Professor Porsche's relationship with VW chairman Heinz Nordhoff since 1948. VW's annual contracts covered most of Weissach's overhead. Paul Hensler planned the campus and the buildings and supervised the project.

▲ The R models used Typ 901/22 engines from the 906 Carrera 6 race car with 1,991cc displacement. These highly developed engines produced 210 horsepower at 8,000 rpm.

▲ Over 96 hours, the car ran through rain and fog, day and night. At the end of four days, it had driven 20,086.08 kilometers, 12,505.38 miles. *Photograph courtesy Porsche Archive*

▲ To reduce weight, the R provided just three instruments, no radio, and no glove box cover. Mechanics drilled holes in countless panels to shave grams and ounces off the car.

YEAR	1967
DESIGNATION	911R
SPECIFICATIONS	
MODEL AVAILABILITY	Coupe
WHEELBASE	2211mm/87.0 inches
LENGTH	4163mm/163.9 inches
WIDTH	1610mm/63.4 inches
HEIGHT	1320mm/52.0 inches
WEIGHT	800kg/1764 pounds
BASE PRICE	Not available
TRACK FRONT	1337mm/52.6 inches
TRACK REAR	1317mm/51.9 inches
WHEELS FRONT	6.0Jx15
WHEELS REAR	7.0Jx15
TIRES FRONT	185-15 radial
TIRES REAR	185-15 radial
CONSTRUCTION	Unitized welded steel
SUSPENSION FRONT	Independent, wishbones, MacPherson struts, longitudinal torsion bars, hydraulic double-action shock absorbers
SUSPENSION REAR	Independent, semi-trailing arms, transverse torsion bars, hydraulic double-action shock absorbers
BRAKES	Discs, 2-piston cast iron fixed calipers
ENGINE TYPE	Horizontally opposed DOHC six-cylinder Typ 901/22
ENGINE DISPLACEMENT	1991cc/121.5CID
BORE AND STROKE	80x66mm/3.15x2.60 inches
HORSEPOWER	210@8000rpm
TORQUE	152lb-ft@6000rpm
COMPRESSION	10.5:1
FUEL DELIVERY	2 Weber 46J carburetors
FINAL DRIVE AXLE RATIO	4.428:1
TOP SPEED	150mph depending on final drive
PRODUCTION	4 prototype, 20 production

1968 911L

For 1968 Porsche could not export the 911S to U.S. markets. With larger intake and exhaust valves, and Weber carburetors tuned slightly rich, the engine did not meet exhaust emissions standards in effect that year. In response, the company created a 911L, for *luxus*, or "luxury." Engineers fitted it with a 130-horsepower DIN (148-horsepower SAE) flat six and included every interior feature of the S (as well as its ventilated disc brakes), but packaged in a base model. For Europe the L slipped in between the 911T and 911S. For U.S. customers, this was the top model, and engineers added the EPA's mandatory air pump. When Porsche closed the books on model year 1968, it had offered 911, 911L, 911S, and 912 coupes and the Targas to its customers. It had assembled a new record of 14,300 cars, known as the A series, and for the first time, six-cylinder production passed the fours. As the company introduced the B series 1969 models, F. A. Porsche received an award at the Paris Auto Salon for "overall aesthetic conception in the creation of a Porsche body."

The B series brought a significant improvement to 911 handling. Engineers shifted the rear wheels 2.25 inches (57mm) farther back. By lengthening the wheelbase (to 89.3 inches/2,268mm) without stretching the body, they improved weight distribution and handling by effectively transferring some weight to the front wheels. This required new rear suspension trailing arms, and they fitted new half-shaft universal joints to accommodate the greater angle from engine to wheels.

Model year 1969 was significant in other ways. Porsche introduced a new Bosch mechanically fuel-injected 911, the 911E, with 140 horsepower DIN (158 SAE) and a green fan shroud. This car assumed the middle ground the 911L had occupied in 1968. The Bosch system also allowed Porsche to do away with

YEAR	**1968**
DESIGNATION	**911L**
SPECIFICATIONS	
MODEL AVAILABILITY	Coupe, Targa
WHEELBASE	2211mm/87.0 inches
LENGTH	4163mm/163.9 inches
WIDTH	1610mm/63.4 inches
HEIGHT	1320mm/52.0 inches
WEIGHT	1080kg/2376 pounds
BASE PRICE	$5,375 coupe – $5,729 Targa
TRACK FRONT	1337mm/52.6 inches
TRACK REAR	1317mm/51.9 inches
WHEELS FRONT	5.5Jx15
WHEELS REAR	5.5Jx15
TIRES FRONT	165HR-15 radial
TIRES REAR	165HR-15 radial
CONSTRUCTION	Unitized welded steel
SUSPENSION FRONT	Independent, wishbones, MacPherson struts, longitudinal torsion bars, hydraulic double-action shock absorbers
SUSPENSION REAR	Independent, semi-trailing arms, transverse torsion bars, hydraulic double-action shock absorbers
BRAKES	Discs, 2-piston cast iron fixed calipers
ENGINE TYPE	Horizontally opposed DOHC six-cylinder Typ 901/06
ENGINE DISPLACEMENT	1991cc/121.5CID
BORE AND STROKE	80x66mm/3.15x2.60 inches
HORSEPOWER	130@6100rpm
TORQUE	128lb-ft@4600rpm
COMPRESSION	9.0:1
FUEL DELIVERY	2 Weber 40 IDA carburetors
FINAL DRIVE AXLE RATIO	4.428:1
TOP SPEED	130mph
PRODUCTION	1,169 coupes; 444 Targas

◀ Emissions regulations stopped Porsche from exporting the 911S model to American customers. The L coupe sold for 21,450DM (roughly $5,376) and $6,790 in the United States.

▲ **1968 911L Coupe**
Porsche expanded its lineup for 1968, adding a 110-horsepower T (for Touring) to replace the base 911, and an L (for Luxus or luxury) below the 160 horsepower S. The Typ 901/06 L engine provided buyers with a 130-horsepower engine.

the air pump on its U.S. models, so the 911S returned to the American lineup, now with 170 horsepower DIN (190 SAE). Porsche introduced cast-aluminum crankcases and discontinued the magnesium castings used since the car's launch.

Engineering introduced a hydropneumatic, self-adjusting front suspension on the 911E. This system, manufactured by Boge, automatically adjusted to front-end loads as subtle as fuel tank level changes or the addition of luggage for a vacation. Porsche engineers designed the system to incorporate the race-proven ATE disc brake calipers. These accommodated pads that were 50 percent larger than Porsche's original steel calipers. Finally, in the interest of better front-end weight distribution, Porsche switched a single 12-volt battery to one 6-volt unit beneath each headlight.

▲ **1969 912 Coupe**
The 912 used Porsche's Typ 616/36 four-cylinder 1,582cc (96.5-cubic-inch) engine. A four-speed transmission was standard. Acceleration to 200 kilometers per hour took 13.5 seconds.

1965-1969 912

Porsche discontinued the 912 at the end of model year 1969. The 914, introduced in September 1969, was the new entry-level Porsche, the task previously assigned to the 912. Zuffenhausen assembly needed the space it had dedicated to 912 production to manufacture the much more potent 914-6. Its older engine, too, confronted new emission standards.

About this time, the company hired Anatole "Tony" Lapine to "understudy" F. A. Porsche as design chief. Lapine was styling boss at Opel and before that had worked on Corvette design at GM in Detroit. In a recent interview, Lapine recalled that Ferry told him F. A. would take over as Porsche managing director upon Ferry's retirement and that Lapine was to assume the top design job. When he arrived in Zuffenhausen, he found the company working on the Typ 914 mid-engine prototypes for a project developed with, but orphaned from, VW. VW chairman Heinz Nordhoff had approved the project in 1967 as a potential sports car for the VW line. But Nordhoff, who was scheduled to retire in December 1970, died in April 1968, and the board's choice as successor, Kurt Lotz, was not a sports car enthusiast, so work reverted to Porsche ownership.

In the meantime, Ferdinand Piëch had started engineers working on the "next" VW Beetle, project EA266, and the next 911, Porsche's Typ 1966. Both companies worried about the longevity of the air-cooled engine concept. F. A. Porsche's first prototype for the 356 successor, the Typ 695, used a flat four-cylinder engine mounted under the rear seat between the rear wheels. Piëch resurrected the concept but took it much further for EA 266 and Typ 1966. His engineers had created the flat 12-cylinder 917 race car engine by fitting together two flat sixes. He challenged them to create a flat eight by doing the same thing, mounting two inline water-cooled

▲ The 912 offered buyers the same elegant body and interior with slightly less weight at the rear. The 912 coupes weighed 950 kilograms (2,090 pounds) compared with the 911T at 1,020 kilograms (2,244 pounds).

flat fours in opposition to each other. He asked Lapine's designers to come up with a body for a mid-engine coupe.

"It was beautiful," Lapine explained decades later. "Like nothing you've ever seen. Mr. Piëch wanted three lines, just like the T, E, and S, but as a four-cylinder, an eight-cylinder, and for special customers and racing, a twelve. He even built a prototype flat eight-cylinder engine. Three seats, driver in the middle ahead of the engine. Large trunk up front, big storage over the engine. This was to be the next 911." Porsche planned to replace the first generation of cars in 1972 or 1973, and Typ 1966 was to appear then.

▲ The 912 coupes sold for 17,538DM ($4,474) at the factory. American buyers paid $5,235. Houndstooth upholstery was optional on 912s in 1969. The owner did the door panels and the dashboard.

YEAR	1965-1969
DESIGNATION	912
SPECIFICATIONS	
MODEL AVAILABILITY	Coupe, Targa introduced in 1967
WHEELBASE	2211mm/87.0 inches – 2268mm/89.3 inches in 1969
LENGTH	4163mm/163.9 inches
WIDTH	1610mm/63.4 inches
HEIGHT	1320mm/52.0 inches
WEIGHT	950kg/2090 pounds
BASE PRICE	$4,395 (1969 coupe)
TRACK FRONT	1337mm/52.6 inches – 1362mm/53.6 inches in 1969
TRACK REAR	1317mm/51.9 inches – 1343mm/52.9 inches in 1969
WHEELS FRONT	5.5Jx15 (in 1969)
WHEELS REAR	5.5Jx15 (in 1969)
TIRES FRONT	165HR-15 radial (optional)
TIRES REAR	165HR-15 radial (optional)
CONSTRUCTION	Unitized welded steel
SUSPENSION FRONT	Independent, wishbones, MacPherson struts, longitudinal torsion bars, hydraulic double-action shock absorbers
SUSPENSION REAR	Independent, semi-trailing arms, transverse torsion bars, hydraulic double-action shock absorbers
BRAKES	Discs, 2-piston cast iron fixed calipers
ENGINE TYPE	Horizontally opposed OHV four-cylinder Typ 616/36
ENGINE DISPLACEMENT	1582cc/96.5CID
BORE AND STROKE	82.5x74mm/3.25x2.92 inches
HORSEPOWER	90@5800rpm
TORQUE	90lb-ft@3500rpm
COMPRESSION	9.3:1
FUEL DELIVERY	2 Solex 40 PII-4 carburetors
FINAL DRIVE AXLE RATIO	4.428:1
TOP SPEED	115mph depending on final drive
PRODUCTION	28,333 coupes, 2,562 Targas (total all years)

▶ Records suggest 35 or fewer of these T/R composites were created. This car finished 5th in Group 3, 12th overall, in the 1968 Monte Carlo Rally.

▼ Swedish rally driver Åke Andersson competed for the Scania Vabis rally team. The car weighed just 923 kilograms, 1,846 pounds.

▲ Two engines were available for the T/R. This was a 235-horsepower version of the 2-liter 901/22 with Weber 48IDA carburetors used in the 906 Carrera 6 race car and in the 911R. The other option was a 180-horsepower 911S engine, earning those cars the name ST.

▲ The cockpit of the 1968 T/R featured only the bare necessities for winning races.

▲ When Porsche introduced the 911T for 1968, it also found a way to utilize parts remaining from the 1967 911R assembly. The T/R was not an actual model, but one racers created by ordering parts and installing them.

▲ Herbert Linge, Dieter Glemser, and Willi Kauhsen shared driving duties for 84 hours over three and a half days. They won by completing 356 laps, for a total of 8,325.7 kilometers, 5,203.5 miles, for an average of 99.1 kilometers (61.95 miles) per hour.

▶ For 1969, Porsche lengthened the wheelbase of the 911 by 57mm (2.24 inches) to 2,268mm (89.3 inches) without changing the car body. This improved handling and balance by shifting some weight to the front wheels.

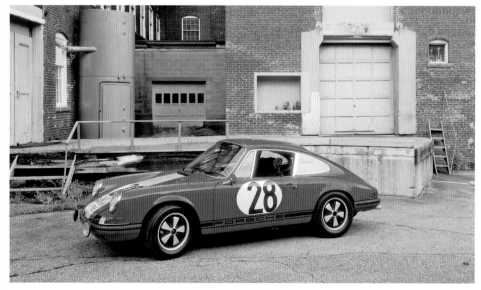

▲ Porsche sent three brand-new 1969-model endurance-race-prepared 911S models to the Marathon de la Route in August 1968. Run for 84 hours nonstop on Nürburgring's *Nordschleife*, the north circuit, each lap was 22.81 kilometers, 14.17 miles.

◀ Some reports and records suggest this car ran—and won—the Marathon de la Route using Porsche's new Sportomatic transmission. But co-driver/winner Herbert Linge confirmed the car ran with a reinforced five-speed manual gearbox.

1970–1971 **911T**
1970–1971 **911E**
1972–1973 **911S**
1973 **CARRERA RS**
1973 **CARRERA 2.8 RSR**
1974 **CARRERA 3.0 RSR**
1974–1975 **911**
1975–1977 **TURBO**
1976 **912 E**
1976–1977 **934 AND 935**

CHAPTER 2
THE FIRST GENERATION CONTINUES 1970–1977

1970-1971 911T

While Porsche's experimental staff worked on the Typ 1966 911 replacement, production engineers and designers moved forward with the C series for model year 1970. Automakers throughout the world struggled to control hydrocarbon emissions, and Paul Hensler's staff designed a new series of 2.2-liter engines to address these needs. They enlarged bore from 80mm to 84, increasing overall displacement from 1,991cc to 2,195 (from 121.5 to 133.9 cubic inches). The black fan–shrouded T now developed 125 horsepower DIN (142 SAE); the green-shrouded E produced 155 horsepower DIN (175 SAE), and the red-shrouded S turned out 180 DIN, 200 SAE horsepower.

In Zuffenhausen, as the company began manufacturing its 1970 models, it completed construction of its new 160,000-square-foot (14,864-square-meter) assembly plant, paint shop, and interior trim facility. Porsche offered nine standard exterior colors, and the new paint shop made custom work possible. Dealers in California often ordered cars in special colors on speculation. Their gambles paid off, and these automobiles attracted high-visibility entertainers and athletes. The company liked seeing these colors as they left the factory; over time, Porsche added them to options catalogs, inviting dealers to submit orders for personalized cars.

The value of the deutsche mark rose during the D series 1971 model year, trading at DM 3.64 to the dollar at the beginning of the calendar year and 3.27 at year-end. This pushed up prices in the United States and slowed demand.

◀ **1970 911 S 2.4 Tour de France**
It was the lightest 911 to that time, weighing 780 kilograms, 1,716 pounds. Porsche quoted the top speed at 243 kilometers (152 miles) per hour.

▲ (Top) The Targa weighed 1,020 kilograms, 2,244 pounds. It sold for 21,911DM ($6,003) at the factory. The company manufactured 2,545 Targas and about 6,544 coupes.

▲ (Bottom) For model year 1970, Porsche introduced a new 2.2-liter 2,195cc (133.9-cubic-inch) engine, the Typ 911/03. This T version developed 125 horsepower at 5,800 rpm.

Zuffenhausen cut production from 70 cars a day to 60. This cut dropped annual assembly figures to their lowest level in five years, 11,715 cars. As the last battalions of engineers and designers moved from Zuffenhausen into their new offices and studios at Weissach, their mood was subdued.

Ferry Porsche was an excellent steward of his company's finances and production. The 1970–1971 price increases American customers experienced resulted from currency exchanges, not poor management. But where Ferry had been cautious, his nephew Ferdinand Piëch had been less so, treating the Racing Department as if it had endless reserves. His 911Rs, 906s, 908s, 909s, 910s, and 917s had won countless races and successive world championships, bringing acclaim to the company.

F. A. Porsche had benefited from being Ferry's son, a position that allowed him to deliver an idea and have it approved. Ferdinand Piëch, as nephew, had always held out for the highest quality pieces and workmanship, even when the difference between best and excellent was minimal but at a cost of five or ten times more. His Racing Department routinely built new race cars from one event to the next, while other competitors overhauled last week's winner. He brought in engineers like Norbert Singer, Tilman Brodbeck, and many others for their expertise in aerodynamics and other specialties.

His 917s in 1969 and 1970 were born of anxiety inside Volkswagen. VW had watched sales of its air-cooled Beetle fall from previous records; it worried about the future appeal of air-cooled engine power. VW offered to underwrite any future racing program Porsche undertook, so long as it continued with air cooling, and it put no ceiling on the

▲ Porsche provided four ventilated disc brakes on the T series. A new top on the Targa was lighter and folded more easily for storage.

costs. The indomitable 917s proved for all time the power and reliability of air-cooled engines, bringing additional championships to Porsche, even as VW spent millions of deutsche marks.

These extraordinary investments, as well as regular arrivals from the Porsche and Piëch families at factory doors, looking for jobs in the family company, put Ferry Porsche and his sister Louise Piëch at increasing odds. The Supervisory Board included 60-year-old Ferry; his son F. A. Porsche, then 34; Ferdinand Piëch, 33; his younger brother Michael, 28; and Porsche's chief financial officer, Heinz Branitzki, 41. F. A.'s younger brother Peter, then 29, had joined the company in 1963. By 1970, while not yet a board member, he was head of production.

In October 1970, Ferry called a meeting at the family compound in Zell am See, Austria. He hoped to broker a peace among the many individuals and ideas he had listened to in the previous year. It did not go well, and within days, members of both families agreed to vacate their jobs within the company and to hire qualified outside professionals to replace them. Ferry alone remained as chairman of the Supervisory Board.

The transition occurred gradually, and it benefited family members financially. However, it transformed the company, even as it buried animosities that were to surface decades later. Porsche evolved from a family-run company into a limited-liability corporation. If that decision was the first to set in place profound effects, the second came when the supervisory board rehired engineer Ernst Fuhrmann as purchasing and production chief.

Fuhrmann had been a design engineer at Porsche in the 1950s. He had created the Typ 547 Carrera four-cam engine. These power plants delivered to Porsche many of its early outright competition victories. When the board passed him over for a promotion he believed he deserved, he left in 1956 to join Goetze, a piston ring and bearing maker. But in early 1971, he quit Goetze in a dispute over management duties.

Piëch and Helmuth Bott visited Fuhrmann at his home in Teufenbach, Austria, and asked him to return to Porsche. "They showed me designs for the new cars," Fuhrmann explained in an interview in 1991. "I had nothing else to do. I should say, the position was very simple and easy to handle."

By September 1971, Fuhrmann was in charge of all things technical at

Weissach. He and the board elevated Bott to head of development, design, and testing. Fuhrmann arrived with work well along on the "next" 911, the three-seat Typ 1966 mid-engine car. A month later, management changes at VW killed the project. Development funds to finish the Typ 1966, set aside out of Weissach's profits on its work for VW, dried up.

"But that was another thing that complicated things," Fuhrmann continued. "The successor to the nine-eleven was already along. The sales people wanted a new car." The board had planned to end 911 production in 1973.

"Overnight everything disappeared," Tony Lapine explained. "The next day it was as if these projects never had existed. All the drawings, the notes, the papers. Mr. Piëch's prototype engine. I heard it was cut up." Weissach had no work. Fuhrmann had no successor to the 911. And the last of the Porsche and Piëch family members left the company. "And then we had a hell of a job to do," Lapine said, "to keep this place going."

In the midst of corporate and family drama, production for 1972 model year cars continued. The 1976 Clean Air Act required automobiles sold in the United States starting in 1976 to emit 90 percent fewer hydrocarbons than 1971 levels. Manufacturers concluded that the only way to achieve these reductions was to use catalytic converters. These devices burned the already hot engine exhaust, baking the unburned emissions before they reached the atmosphere. Tetraethyl lead, the gasoline additive that increased octane ratings to provide higher performance, killed the heating elements within converters. In the early

▲ **1970 911 ST Monte Carlo Rally**
Björn Waldegård and co-driver Lars Helmers drove this car to first overall in the 1970 Monte Carlo Rally. It was their second win and the third overall for Porsche's 911.

▶ **1971 911E Coupe**
Porsche introduced the 911 E series in the middle of the line up for model year 1969. Originally featuring a 2.0-liter injected engine, Porsche enlarged it to 2.2 liters for 1970. It developed 155 horsepower (DIN) at 6,200 rpm.

1970s, automakers began redesigning engines to operate on lead-free fuels.

Porsche engineers recognized that their short-stroke, large-bore, high-rpm engines developed substantial horsepower but produced dirtier exhaust. Small bores coupled to longer strokes combusted fuels more fully, meeting new emission standards more easily. They also provided greater torque at lower engine speeds, a help for Porsche drivers confronting stop-and-go traffic.

Porsche kept cylinder bore at 84mm but lengthened stroke from 66mm to 70.4mm. This brought overall displacement to 2,341cc, or 142.8 cubic inches. Marketing rounded this figure up to 2.4 liters. To simplify manufacture and to reduce costs, Paul Hensler's engineers designed a single aluminum cylinder head with one camshaft for T, E, and S engines. T models outside the United States retained their Zenith-Solex carburetors, fitted since introduction in 1968. However, engineers fuel-injected U.S.-bound T models, as well as all E and S versions. Monitoring cost effectiveness, Porsche carried over the same engines through the 1973 model year

The 911T now developed 140 horsepower DIN (157 SAE). Hensler's engineers coaxed 165 horsepower DIN (185 SAE) from the E. The S, with 190 horsepower DIN (210 SAE), remained the racer's favorite. On average, these figures represented a 10 percent increase over what 2.2-liter engines had produced. This was a noticeable difference for city drivers. The 2,341cc engine was one of Ferdinand Piëch's last three production car contributions. Second was a new transaxle; better acceleration had been his design, engineering, and performance target. The new gear ratios sacrificed top speed and slightly increased fuel consumption, but they made the cars more enjoyable and easier to drive for most buyers.

This sort of development was typical of Piëch's near obsession with high standards of performance from employees and machines. Historian Karl Ludvigsen, who has thoroughly documented Porsche's history, expressed it best: This new transaxle "reflected an almost reckless drive for perfection in Porsche cars, for only two years earlier the change had been made to a magnesium die-casting for the housing of the earlier trans-axle." Piëch's new transaxle added 20 pounds (9 kilograms) to the car's weight. But he always was concerned about weight at the car's extremes. For the 1972 model, he relocated the dry sump oil reservoir behind the passenger door, ahead of the rear wheel. Body engineers installed a filler cap and lid into the rear fender just behind the B-pillar.

Unfortunately, this external filler confused some innocent gas station attendants who did not know Porsche's fuel filler was still in the trunk. They pumped gasoline into the oil reservoir. For 1973, recognizing the confusion and concerned that U.S. regulations for side-impact safety might restrict importation, engineers returned the tank to the rear fender behind the engine.

For 1972 the sales staff made the Boge hydro-pneumatic struts optional equipment on the 911E. Engineers had learned the struts leaked, and owners discovered they were more costly to replace than standard units or even the expensive Konis. Deleting the Boges reduced the 1972 911E price by DM 300, from 26,680 in 1971 to 26,380, or $7,995 in the United States. The T settled at DM 23,480, or $7,367 in America, and the S reached DM 31,180, or $9,495. Porsche charged an additional DM 1,000, $735, for the Targa on any platform.

Despite its slightly lower top speed, 140 miles per hour (225 kilometers per hour), the S still was a potent automobile, accelerating from 0 to 100 miles per hour (160 kilometers per hour) in 15 seconds. Wider tires and fenders increased the frontal area from 17.4 square feet at introduction in 1964 to 18.4 square feet for the 1972 S, raising the coefficient of drag from 0.38 to 0.41. At high speeds, air pressure tended to lift the nose of the car.

YEAR	**1970-1971**
DESIGNATION	**911T**
SPECIFICATIONS	
MODEL AVAILABILITY	Coupe, Targa
WHEELBASE	2268mm/89.3 inches
LENGTH	4163mm/163.9 inches
WIDTH	1610mm/63.4 inches
HEIGHT	1320mm/52.0 inches
WEIGHT	1020kg/2244 pounds
BASE PRICE	$5,471 coupe - $6,003 Targa
TRACK FRONT	1362mm/53.6 inches
TRACK REAR	1343mm/52.9 inches
WHEELS FRONT	5.5Jx15
WHEELS REAR	5.5Jx15
TIRES FRONT	165HR-15 radial
TIRES REAR	165HR-15 radial
CONSTRUCTION	Unitized welded steel
SUSPENSION FRONT	Independent, wishbones, MacPherson struts, longitudinal torsion bars, hydraulic double-action shock absorbers
SUSPENSION REAR	Independent, semi-trailing arms, transverse torsion bars, hydraulic double-action shock absorbers
BRAKES	Discs, 2-piston cast iron fixed calipers
ENGINE TYPE	Horizontally opposed DOHC six-cylinder Typ 911/03
ENGINE DISPLACEMENT	2195cc/133.9CID
BORE AND STROKE	84x66mm/3.31x2.60 inches
HORSEPOWER	125@5800rpm
TORQUE	130lb-ft@4200rpm
COMPRESSION	8.6:1
FUEL DELIVERY	2 Weber 40 IDT 3C carburetors
FINAL DRIVE AXLE RATIO	4.428:1
TOP SPEED	127mph
PRODUCTION	11,019 coupes; 6000 Targas

▲ The coupe sold for $7,607 (26,473DM). Records suggest the company manufactured 3,028 coupes from 1969 through 1971.

1972–1973 911S

Tilman Brodbeck, a young racing engineer, joined Porsche in October 1970. He had earned a technical degree in mechanical engineering and aerodynamics. Early in 1971, Helmuth Bott assigned him to work on the 911S front-end lift in Stuttgart University's wind tunnel. Design chief Tony Lapine sent one of his modelers to join Brodbeck, in case the engineer got any ideas.

A technician in the wind tunnel moved an oil vaporizer wand over the front end of the car. Brodbeck watched the oil stream disappear directly under the front valence below the bumper. Wondering how to stop it or redirect it, he stuck his finger into the path of the oil stream. The air detoured around it. They scrounged around the facility for other things that might deflect that stream. By the end of the first day, Brodbeck had taped a piece of rope across the bottom lip. The oil stream caught on the rope and swerved left or right. The control room technician read encouraging results. By the end of the third day, they had roughed in a smooth fiberglass piece that had an even greater effect. It reduced lift from 183 pounds to 102 and dropped the drag coefficient from 0.41 to 0.40. Bott gave the project high priority to develop and introduce this spoiler as a production part and as an accessory that owners could mount on earlier cars.

The E series 1972 models gave way to F series 1973 models. The company earned revenues of slightly more than DM 300 million (roughly $94 million) in 1972, and its motorsports efforts had cost more than DM 30 million ($9.4 million). The economic downturn in the United States in 1971 and 1972 reduced Porsche exports, decreasing revenues anticipated for future projects and planned models. During this time,

▲ **1972 911S Targa**
Porsche sold the Targa for 32,900DM, $10,313 at the time. It weighed 1,075 kilograms, 2,365 pounds.

▲ Porsche hurried the front spoiler into production in time for 1972 models. The company introduced a new 2,342cc (142.8-cubic-inch) Typ 911/53 engine for the 1972 S. It developed 190 horsepower at 6,500 rpm.

the phased withdrawal of Porsche and Piëch family members matched the numbers of new engineers, stylists, and managers arriving almost weekly. The company was in transition, its resources were limited, and few at Porsche had a clear view of the future.

The man in charge was a performance enthusiast. Ernst Fuhrmann admitted that one of his motivations for conceiving four-cam heads and designing the Typ 547 Carrera engine was to obtain something fast to power his company 356. In early 1972, he went to Hockenheimring for a touring car race with other Porsche engineers.

"I was in the pits," Fuhrmann recalled during an interview nearly 20 years later.

"I watched many nine-elevens. Fords and the BMWs were passing them. Even our fastest nine-eleven was lapped by a Ford and a BMW. I went looking for one of our engineers, Mr. [Norbert] Singer, to ask him why this happened. I found another one, younger, one of Mr. Singer's protégés."

Singer's protégé was Wolfgang Berger, who told Fuhrmann that Ford had developed a small series of *rennsport* models, racing prototypes in effect, that took advantage of every loophole in FIA racing regulations. Berger went on, volunteering what Porsche might need to win against them.

"Your analysis is interesting," Fuhrmann recalled telling him. "Think about it and then tell me what you will do." Berger reported the conversation to Singer, who passed along the information to Bott. Within days of the race, Bott summoned Tilman Brodbeck back into his office.

"Porsche nine-eleven race drivers have a lot of trouble when they go for the curves on the racetrack," Bott explained. "Even the Ford Capris and the BMW coupes are quicker through the curves. You must do something. Anything! Without changing the whole car, it must be possible that these racers can buy something to make the car better. Think about it!"

"We went back into the Stuttgart University wind tunnel," Brodbeck

▲ The S engine accelerated the Targa and coupe from 0 to 100 kilometers per hour in 7.0 seconds. Top speed was 230 kilometers (140 miles) per hour.

recalled, "the same one where we had done the front lip for the Carrera. Now we started with welding wire to make a form. The front had needed such a little bit. We thought about how to change the shape of the rear. Over the next two or three days, we formed this new shape. We moved it forward and back, up and down the rear end. It was trial and error."

Brodbeck's final version was flat sheet metal wrapped over welding wire. His colleagues in engineering laughed when they saw it mounted on the back of a car, but Brodbeck remembered it was nervous, stress-relieving laughter. Günther Steckkönig, a Porsche test driver since 1953 and a factory race driver by the mid-1960s, drove it around Weissach's test track. He came in and said, "Wait, the car feels much better." The project went from Weissach's test track to Tony Lapine's design studios, where Brodbeck asked them to make something smoother and better looking without changing the aerodynamics. "We told them the important points for the aerodynamics, where they have to be. Then they made this little thing, this ducktail, the bürzel." Then it went back into the wind tunnel, this time to Volkswagen's, since Mercedes-Benz had just acquired the Stuttgart University facility. A former classmate of Brodbeck's was working at VW in aerodynamics. Hearing that Tilman was there to "manage the lift," he said that Porsche was wasting its time. "The only thing we are working on is the drag coefficient. That's the most important thing."

Shortly after that, Bott took the mocked-up prototype to VW's test track at Ehra-Lessien. Bott had a series of handling exercises that could unsettle a car. One involved a violent lane-change maneuver. Acceptable chassis and suspension development allowed just three fishtail swerves for the car to stabilize or it went back for more work. Bott invited Brodbeck along for the first ride.

"He was driving about one-hundred eighty kilometers [roughly 112 miles per hour] on a straight," Brodbeck explained. Bott suddenly yanked the steering wheel hard to the right. "You can imagine what that does. I got pale. Without the spoilers it was awful what this car did. And with the spoilers it was amazing. Everybody said, 'Well, something else must be changed, the tires, the suspension. It cannot just be these two small spoilers.'" For the tests, Bott had used a single 911, so everyone could watch his mechanics exchange only body panels.

Wolfgang Berger, Singer's protégé, had kept Fuhrmann informed of progress. Berger had direct access, bypassing normal chain-of-command reporting hierarchy. Inspired by lightweight 911R models, Berger gutted a 911T in preparation for what other engineers were doing to meet Fuhrmann's challenge

In the engine, the Biral cylinders used on production 911s had reached near-maximum bore at 88mm. Much more and the cylinder walls got too thin for reliability. Helmut Flegl and Hans Mezger had specified Nikasil, a nickel-silicon carbide, to line the engine cylinders on 917s, and this allowed a few extra millimeters. Mezger increased the production 911 bore to 90mm from 88, which enlarged total displacement to 2,687cc, or 163.9 cubic inches. This advanced the 911 to the next racing class, Under 3-Liters.

The FIA's "production" classifications allowed entrants to run wider wheels. This encouraged Berger to enlarge rear fenders. Brakes from the 917s fit in these new wheels, and these ran a prototype ABS system from Teldix. Fuhrmann asked Paul Ernst Strähle, a longtime Porsche racer and company friend, to enter the car as his own in the Group 5 prototype category for the 1,000-kilometer race in Austria. Berger fitted the front end with a special spoiler derived from what Peter Falk, Norbert Singer, and racer Mark Donahue had devised for Donahue's Can-Am 917-30. This incorporated an oil cooler even as it channeled airflow off the front. Berger mounted a large version of Brodbeck's ducktail, which reached the height of the rear window. Fuhrmann hoped that using Strähle as the entrant might avert suspicion that the car was a factory prototype. It finished in 10th overall, right behind sports racing prototypes. It signaled to Porsche's competitors that the 911 no longer was an ill-handling loser.

It was a significant test for Fuhrmann. He had inherited a company that planned to drop the 911. He had neither a replacement nor money to create something new. He had to reinvigorate the existing car and somehow make it exciting to buyers. Production racing could do that—a sales technique that emphasized the character of this new racer as Fuhrmann's Carrera engines had done for 356 buyers.

Fuhrmann assigned Norbert Singer to manage competition development and racing programs. Wolfgang Berger's production engineering staff and Tony Lapine's styling studio collaborated with Singer on the new 911 S 2.7, as it was to be known. The goal was to manufacture a Group 4-legal "production" model, stripped of anything that might jeopardize racing capabilities yet still appealing to 500 paying customers. Fuhrmann considered it an addition to the 911 lineup. It offered better performance than the 2.4-liter 911S already in production.

YEAR	1972-1973
DESIGNATION	911S
SPECIFICATIONS	
MODEL AVAILABILITY	Coupe, Targa
WHEELBASE	2271mm/89.4 inches
LENGTH	4147mm/163.3 inches
WIDTH	1610mm/63.4 inches
HEIGHT	1320mm/52.0 inches
WEIGHT	1075kg/2365 pounds
BASE PRICE	$9,655 coupe - $10,313 Targa
TRACK FRONT	1362mm/53.6 inches
TRACK REAR	1343mm/52.9 inches
WHEELS FRONT	6.0Jx15
WHEELS REAR	6.0Jx15
TIRES FRONT	185/70VR15
TIRES REAR	185/70VR15
CONSTRUCTION	Unitized welded steel
SUSPENSION FRONT	Independent, wishbones, MacPherson struts, longitudinal torsion bars, hydraulic double-action shock absorbers
SUSPENSION REAR	Independent, semi-trailing arms, transverse torsion bars, hydraulic double-action shock absorbers
BRAKES	Discs, 2-piston aluminum-front, cast iron-rear fixed calipers
ENGINE TYPE	Horizontally opposed DOHC six-cylinder Typ 911/53
ENGINE DISPLACEMENT	2341cc/142.9CID
BORE AND STROKE	84x70.4mm/3.31x2.77 inches
HORSEPOWER	190@6500rpm
TORQUE	159lb-ft@5200rpm
COMPRESSION	8.5:1
FUEL DELIVERY	Bosch mechanical fuel injection
FINAL DRIVE AXLE RATIO	4.428:1
TOP SPEED	140mph
PRODUCTION	3,160 coupes; 1,894 Targas both years

▲ This was Porsche's first right-hand-drive specification RS Carrera lightweight delivered to the United Kingdom. The company went on to assemble a total of 69 right hand drove RS Carreras.

1973 CARRERA RS

Those in marketing and sales who had worried the 911R into oblivion saw a repeat performance coming, and they raised obstacles as they ran away. Fuhrmann remained firm: What if Zuffenhausen assembled 500 cars stripped as Singer needed them for homologation? However, buyers could check an option code giving them the luxurious 911S interior with soundproofing and steel bumpers. Marketing objected to the name S 2.7. Yet they had already discussed words that revisited Porsche history for future model designations. They named this new car Carrera and added RS, stretched between the wheels in a color matching the painted five-spoke wheels from Fuchs.

Engineers, designers, and marketing teams devised the second trim and interior option level. Norbert Singer's homologation version was Porsche's most Spartan configuration of the car. The company delivered the car, designated the RS H (for "homologation"), with narrow wheels and no anti-sway bars, knowing that racers mounted their own equipment. The next level was the lightweight sports version, option M471, with wider wheels and thick anti-sway bars. For those enamored of the style of the new Carrera RS as a race car but not wanting to sacrifice creature comforts, a third level, option M472, delivered a car with interior appointments similar to 911S production models. The fourth configuration, M491, was available to known racers, providing them a 2.8-liter racing engine, roll bar, 11-inch-wide rear wheels, and flared rear fenders. This was the RSR, or *rennsport rennen* version, Porsche's pure competition model.

The first production prototype appeared in April 1972. Wolfgang Berger started from a 1972 E series 911T body that still bore its external oil filler cap on the passenger side. It was complete except for front chin and rear ducktail spoilers and the Carrera graphic. Eight more prototypes rolled out of Weissach before the production run set at 500 copies commenced in October. On Fuhrmann's orders, sales and marketing jumped to work. They had a single prototype to show, but they left piles of brochures at interested distributors and dealers. Some previous 911S buyers received personal visits.

Marketing set the Paris Auto Salon as the public debut. By the time doors opened on October 5, 1972, Porsche had 51 orders for the M471 Sport version. The company established a price of DM 33,000 ($11,785 at the time) for this package. An additional DM 2,500 (roughly $893) bought the M472 "touring" package with the 911S interior. Porsche sold the 500 it needed for homologation within a week of the end of the Paris show. Fuhrmann shrewdly let production run until demand ebbed. That happened in July 1973, when number 1,580 drove off the Zuffenhausen assembly line. Porsche increased the price by DM 1,000 (about $330) after the first 500 sold. But no one complained, and orders continued steadily. The additional production reclassified the Carrera as a Group 3 touring car, where it virtually owned the races.

Porsche intended the Carrera RS for FIA racing events contested throughout Europe. To be competitive, engineers needed to run the engine without the emissions controls required in the United States, so Porsche made no effort to "federalize" the RS 2.7-liter engine. They were unobtainable to American buyers for years, until the EPA relaxed standards for older cars, enabling enthusiasts and collectors to import and enjoy them.

Demand for the car pleased the sales staff based in Ludwigsburg, and it taught them a valuable lesson: Porsche owners were devoted to the marque. Something unique was desirable, and they were willing to pay to be part of a small group who could get a Carrera RS.

It was all about exclusivity, and while the company had just recognized it as a marketable option for outsiders, it had been creating special cars for insiders since the beginning. By 1973 Customer Service, Sonderwunsch (Special Wishes), and other departments had produced a dozen or more unique 356s and 911s for Porsche and Piëch family members. Out of the run of 1973 RS Carreras, a special car for Ferry Porsche, painted Silver Green Diamond Metallic, appeared. According to RS Carrera historian Georg Konradsheim, the car boasted a blue cloth-and-leather interior and, in a hint of styling director Tony Lapine's ultimate plans, body color or flat black trim. The side Carrera logo was silk-screened in green. Instead of the Carrera's signature Fuchs wheels, this car had ATS "cookie cutters," also in green. Under the rear deck lid, the car ran with a 2.8-liter engine with a forerunner of Bosch's K-Jetronic continuous fuel injection, which appeared on series production cars half a year later. Ferry drove the car regularly until 1975. It disappeared for decades but ended up in the significant collection that historian Konradsheim has built outside Vienna, Austria.

The absolute truth of value recognized and value paid for held for competitors too. Norbert Singer's Racing Department completed 57 RSR 2.8 models. Porsche charged DM 59,000 for the race car, $22,500 delivered in the United States (or about DM 25,000—$8,930— more than the lightweight RS production model). To create the cars, Singer's mechanics steered RS H homologation cars off Zuffenhausen assembly lines without engines or transmissions. They rolled them from the assembly plant to the former racing shops in Werke I, an area designated for customer service after the Competition Department had moved into Weissach. Once inside the old brick building, mechanics installed engines that Hans Mezger had bored an extra 2mm to 92mm, which gave total displacement of 2,806cc, or 171.2 cubic inches. Mezger's engineers installed cylinder heads that increased compression, used larger valves, and used a twin spark plug ignition. Porsche rated these engines at 300 horsepower DIN (about 286 SAE net) in a car weighing 852 kilograms, or 1,875 pounds.

The 2.8 RSR racing debut took place at the Tour de Corse in November 1972, but things didn't go well. Singer, Mezger, and their staffs worked throughout the winter to correct problems. Martini & Rossi signed a three-year sponsorship contract. The 1973 season began at Daytona, Florida, in early February, toured Europe through the spring and early summer, and concluded in late July at Watkins Glen, New York.

▲ Porsche offered the RS Carrera in 13 standard and 16 special-order colors. The company assembled just 62 in black. Curiously, records suggest most black cars were shipped to Central and South America.

▶ Touring versions accelerated from 0 to 100 kilometers in 6.3 seconds, compared with the lightweight cars that reached 100 in 5.8 seconds. Both models sold for 34,000DM, $12,830 at the time.

◀ (Opposite) The new 2.7 liter 2,687cc (163.9-cubic-inch) Typ 911/83 flat six used Bosch mechanical fuel injection to help reach output of 210 horsepower at 6,300 rpm. One of the car's most outstanding exterior details was its *burzel*, or ducktail spoiler.

▲ (Top) Deleting the clock from the fifth instrument pod saved grams and ounces, as did removing the glove box cover and using a leather pull strap to open the door from the inside. Porsche listed weight at 960 kilograms, 2,112 pounds.

▲ (Bottom) Touring versions weighed 1,075 kilograms (2,365 pounds), which was 115 kilograms (253 pounds) more than the lightweight models. These received inner door panels, glove box covers, radios, clocks, sun visors, and reclining seats.

YEAR	**1973**
DESIGNATION	**911 Carrera RS**
SPECIFICATIONS	
MODEL AVAILABILITY	Coupe: Lightweight M471; Touring M472
WHEELBASE	2271mm/89.4 inches
LENGTH	4147mm/163.3 inches
WIDTH	1652mm/65.0 inches
HEIGHT	1320mm/52.0 inches
WEIGHT	920kg/2024 pounds (M471 Lightweight)
	1075kg/2365 pounds (M472 Touring)
BASE PRICE	$13,094 Lightweight - $13,774 Touring
TRACK FRONT	1372mm/54.0 inches
TRACK REAR	1394mm/54.9 inches
WHEELS FRONT	6.0Jx15
WHEELS REAR	7.0Jx15
TIRES FRONT	185/70VR15
TIRES REAR	215/60VR15
CONSTRUCTION	Unitized welded steel
SUSPENSION FRONT	Independent, wishbones, MacPherson struts, longitudinal torsion bars, hydraulic double-action shock absorbers
SUSPENSION REAR	Independent, semi-trailing arms, transverse torsion bars, hydraulic double-action shock absorbers
BRAKES	Discs, 2-piston aluminum-front cast iron-rear fixed calipers
ENGINE TYPE	Horizontally opposed DOHC six-cylinder Typ 911/83
ENGINE DISPLACEMENT	2687cc/163.5CID
BORE AND STROKE	90x70.4mm/3.54x2.77 inches
HORSEPOWER	210@6300rpm
TORQUE	188lb-ft@5100rpm
COMPRESSION	8.5:1
FUEL DELIVERY	Bosch mechanical fuel injection
FINAL DRIVE AXLE RATIO	4.429:1
TOP SPEED	152mph M471 Lightweight - 149mph M472 Touring
PRODUCTION	17 homologation; 200 M471 Lightweight; 1,308 M472 Touring

1973 CARRERA 2.8 RSR

During the April trials for Le Mans, Norbert Singer ran the cars with 11- and 12-inch wheels. Shortly afterward, loopholes in regulations—and a well-timed protest from a competitor—allowed Singer to run the RSR as a prototype. This gave him considerable freedom to develop new tricks. As a prototype for the 24-hour race in June, Singer fitted the cars with 15-inch rear tires.

Wider tires, wheel flares, and a huge rear wing slowed the cars' top speeds, to Ernst Fuhrmann's chagrin. Yet these same changes improved cornering, so lap times didn't change and the wider tires lasted longer. Still Fuhrmann was irked. How was it that he, an engine man, gave Singer more power and still the car was slower on the straights? "Okay, tell the drivers we race flat out from the first lap. Twenty-four hours," Fuhrmann ordered. Singer predicted a short race. But Porsche's bulletproof engines and chassis ran a near sprint and still finished 4th, 8th, and 10th overall. By the end of the season, the production-derived 911 Carrera RSR 2.8 had won the manufacturers and drivers championships, beating pure sports prototypes from half a dozen competitors. Series production reached 14,714 cars, a near record. Ernst Fuhrmann, Norbert Singer, Wolfgang Berger, and a host of Weissach engineers and designers breathed life back into a car their predecessors had sentenced to death. What few people outside of Weissach knew was what else the engineers had in mind.

"Racing must have a connection to the normal automobile," Fuhrmann explained in 1991. When he rejoined Porsche, its turbocharged Can-Am and InterSeries racers had built a legend of flawless performance and relentless victory. "So I said to my people, why don't we put this success into our series production cars?"

There had been attempts earlier. In 1969 Valentin Schäffer had mounted turbochargers onto a couple of the 2-liter engines. He installed one in a 911 and the other in a 914/6. Because engine compartment space was tight in each vehicle, the turbo hung outside the body of the 911 and developed serious heat dissipation problems on the 914/6. Long tubes required to channel exhaust to spin the turbines, coupled to more long tubes necessary to get the pressurized fuel mix into the engines, caused tremendous turbo lag. Schäffer assembled another prototype and had it running in April 1973. It still suffered turbo lag, and the 911 chassis struggled to handle 250 brake horsepower. Fuhrmann told his engineers not only that it had to work but that it had to function with the new Bosch K-Jetronic electronic fuel injection that Paul Hensler's engineers had coming on the midyear 911T.

The Bosch system appeared through the entire 911 lineup for 1974. Schäffer made it work on his turbocharged prototypes by bringing air through the K-Jetronic, upward past its metering valve, and then into the compressor side of the turbo. This forced the condensed fuel-air mix to the throttle valve above the engine's new cast-aluminum intake manifold. He and Hans Mezger passed the project along to Herbert Ampferer and Robert Pindar to make it production ready, a process with its own challenges.

Race drivers keep their feet hard on the gas pedal until the instant they brake. They drive through the same turns and straights repeatedly, and they develop a rhythm to keep the turbocharger spinning fast for instant response. Sometimes this involves keeping a foot on the gas pedal while braking. Racers don't coast, and they don't cruise at a steady speed limit. The "partial throttle" condition in production cars is one that allows the turbo to slow down. When the driver needs power and floors the gas pedal, the engine horsepower may double within seconds, but only after a delay. Modulating these dramatic differences required months of experiments, tests, adjustments, and more tests.

▲ Porsche engineer Jürgen Barth co-drove this car with Georg Loos to finish 10th overall at Le Mans in June 1973. Together they averaged 177 kilometers (111 miles) per hour to cover 4,249 kilometers, 2,656 miles.

At the Frankfurt show in September 1973, Porsche showed the results, introducing its production Turbo Coupe to an incredulous audience. Weissach designers and engineers created a silver coupe with RSR flared front and rear fenders that housed wide tires. They finished the car with a large flat rear wing and graphics that stretched the word "Turbo" from the top of the rear wheel arches to the taillights. Motor show enthusiasts who had watched the 917 Can-Am and InterSeries cars race found this new automobile irresistible. Printed materials boasted that Porsche had pulled 280 DIN horsepower (about 267 SAE net) out of its 2.7-liter engine. The car had race car looks and promised race car performance, and regular customers could buy it.

▲ Porsche assembled something like 57 of these RSR models. While the front wheels were 9 inches wide, the rears were 11. They weighed 900 kilograms, 1,980 pounds.

▲ Known inside Porsche as Option M492, the car came with brakes from the 917 race cars, a roll bar, flared fenders, and a $24,000 price tag to competitors in the United States.

▲ Engineers increased bore from 84mm (3.31 inches) to 90mm (3.54 inches), which enlarged overall displacement to 2,806cc (171.2 cubic inches). This Typ 911/72 engine developed 300 horsepower at 8,000 rpm.

YEAR	**1973**
DESIGNATION	**911 Carrera 2.8 RSR**
SPECIFICATIONS	
MODEL AVAILABILITY	Coupe
WHEELBASE	2271mm/89.4 inches
LENGTH	4147mm/163.3 inches
WIDTH	1680mm/66.1 inches
HEIGHT	1320mm/52.0 inches
WEIGHT	900kg/1980 pounds
BASE PRICE	Not available
TRACK FRONT	1402mm/55.2 inches
TRACK REAR	1421mm/55.9 inches
WHEELS FRONT	9.0Jx15
WHEELS REAR	11.0Jx15
TIRES FRONT	230/600-15
TIRES REAR	260/600-15
CONSTRUCTION	Unitized welded steel
SUSPENSION FRONT	Independent, wishbones, MacPherson struts, longitudinal torsion bars, hydraulic double-action shock absorbers
SUSPENSION REAR	Independent, semi-trailing arms, transverse torsion bars, hydraulic double-action shock absorbers
BRAKES	Discs, perforated, with aluminum alloy calipers
ENGINE TYPE	Horizontally opposed DOHC six-cylinder Typ 911/72
ENGINE DISPLACEMENT	2806cc/171.2CID
BORE AND STROKE	92x70.4mm/3.62x2.77 inches
HORSEPOWER	300@8000rpm
TORQUE	217lb-ft@6500rpm
COMPRESSION	10.3:1
FUEL DELIVERY	Bosch mechanical fuel injection
FINAL DRIVE AXLE RATIO	Depends on circuit
TOP SPEED	Depends on final drive
PRODUCTION	57

1974 CARRERA 3.0 RSR

The Carrera served as the foundation of the next generation of Porsche's actual race cars as well. FIA regulations restricted competitors in the Manufacturers Championship to a "silhouette formula"; while modifications to the engine, brakes, transmission, suspension, and even interior and chassis were allowed, the car had to look like one available at a dealership. At circuits around the world, Porsche's RSR 2.8- and 3.0-liter cars won events, captured season titles, and inspired entire racing series. In the United States, Roger Penske, who had run Porsche's 917-30 Can-Am program in 1972 and 1973, ordered 15 identically equipped RSRs for a series he and Riverside Raceway owner Les Richter named the International Race of Champions (IROC). Four events challenged a collection of drivers invited from sports car racing, NASCAR stock cars, and Indianapolis open-wheel racing. Norbert Singer and Ernst Fuhrmann created the IROC cars by blending pieces from RS 3.0 bodies and RSR 3.0 engineering, all wrapped in a wild array of colors. The bodies, shipped to the United States with *bürzel* ducktails, ran the races with Porsche's new "whale tail" flat rear wing and bore an intentionally strong resemblance to series production cars. It worked to entice Sunday race spectators to become Monday dealership customers. Buyers couldn't get the IROC 316-horsepower engine, and they may not have desired the Day-Glo colors, but they could buy something close.

YEAR	1974
DESIGNATION	911 Carrera 3.0 RSR
SPECIFICATIONS	
MODEL AVAILABILITY	Coupe
WHEELBASE	2271mm/89.4 inches
LENGTH	4235mm/166.7 inches
WIDTH	1680mm/66.1 inches
HEIGHT	1320mm/52.0 inches
WEIGHT	900kg/1980 pounds
BASE PRICE	Not available
TRACK FRONT	1472mm/57.6 inches
TRACK REAR	1528mm/60.2 inches
WHEELS FRONT	9.0Jx15
WHEELS REAR	11.0Jx15
TIRES FRONT	230/600-15
TIRES REAR	260/600-15
CONSTRUCTION	Unitized welded steel
SUSPENSION FRONT	Independent, wishbones, MacPherson struts, longitudinal torsion bars, hydraulic double-action shock absorbers
SUSPENSION REAR	Independent, semi-trailing arms, transverse torsion bars, hydraulic double-action shock absorbers
BRAKES	Discs, perforated, with aluminum alloy calipers
ENGINE TYPE	Horizontally opposed DOHC six-cylinder Typ 911/75
ENGINE DISPLACEMENT	2994cc/182.7CID
BORE AND STROKE	95x70.4mm/3.74x2.77 inches
HORSEPOWER	330@8000rpm
TORQUE	231lb-ft@6500rpm
COMPRESSION	10.3:1
FUEL DELIVERY	Bosch mechanical fuel injection
FINAL DRIVE AXLE RATIO	Depends on circuit
TOP SPEED	Depends on final drive
PRODUCTION	69 (including 15 for IROC Series US)

1974–1975 911

Porsche simplified its G series model lineup for 1974. The company dropped the T and E models, offering instead a single 150-horsepower DIN (143 SAE net) coupe or Targa 911 as a base model. This sold for DM 27,000 in Germany, $9,950 in the United States. Next up the scale was the 911S with 175 horsepower (167 SAE net). It went for DM 31,000 and $11,875. European markets could get a 210-horsepower DIN Carrera model with the same engine as 1973 models. Porsche fitted the 167-horsepower S engine to Carreras bound for America's shores to meet emissions regulations. The company priced the Carrera at DM 38,000, or $13,575.

Porsche began the model year using Nikasil for cylinder liners. Midyear, after much testing and development work, the company switched to a new aluminum alloy, 390 aluminum-silicon, or Alusil. This material allowed Porsche to cast a cylinder block without inserting iron cylinder liners. Zuffenhausen engine builders installed Ferrocoated pistons in these new blocks.

The factory delivered European Carrera models with a five-speed transmission as standard equipment. The S and base 911 (and the Carrera for U.S. buyers) ran the four-speed gearbox. Weissach engineers desensitized the Sportomatic shift lever; early owners complained that if they accidentally touched it, it inadvertently disengaged. Porsche set a price of $425 for the new version and offered the five-speed for $250.

Besides meeting ever more stringent emissions standards, the G series 1974 models also had to accommodate U.S. Department of Transportation (DOT) edicts that they could hit a barrier at 5 miles per hour (8 kilometers per hour) and suffer no damage. Porsche designer

▲ U.S. regulations for the 1974 model year required front bumpers capable of withstanding a 5-mile- (8-kilometer-) per-hour impact without damage. Porsche dropped T and E models and produced a base 911, 911S, and 911 Carrera.

▼ Porsche introduced these cast alloy "cookie cutter" wheels made by ATS and painted silver as standard on the 911E and optional on the T for 1973. The company continued the wheels on base 911s in 1974.

▲ This base 911 model ran with Porsche's 2.7-liter, 2,687cc (163.5-cubic-inch) Typ 911/92 engine developing 150 horsepower. The standard transmission had four speeds, but a five-speed gearbox was an option.

▲ To meet U.S. emissions specifications, Porsche introduced new fuel injection with Bosch's continuous injection K-Jetronic system. The 2.4-liter engine developed 140 horsepower. The 911T Targa sold for 25,700DM, or $8,695 in the United States.

▶ (Opposite) **1975 911S Targa**
The S Targa sold for 35,450DM, about $14,410 at the time. The Typ 911/93 engine developed 175 horsepower at 5,800 rpm and accelerated the car from 0 to 100 kilometers in
7.6 seconds.

▲ The Targa bar on 911S (and base editions in Europe) remained brushed aluminum. For the Carrera, it was flat black.

▶ (Opposite) Porsche dropped the base 911 from U.S. distribution, leaving only the 911S and Carrera models. Lime green was a standard exterior color.

Wolfgang Möbius created bumpers that met DOT's requirement without destroying the looks of the 911. These made no effort to hide their function; accordionlike covers hid compression bars fixed to the chassis. Möbius, along with stylist Dick Soderberg and chief studio modeler Peter Reisinger had come to Porsche with Tony Lapine from Opel when he joined the company in 1969.

One of Lapine's influences was to tone down trim materials on Porsche's cars. He gradually evolved 1960s tastes into those of the 1980s. As they had done with the racing RSR 2.8 and 3.0 models, they blackened window frames, outside mirrors, and door handles of the series production cars. Chrome remained optional. Inside the cars, engineers and stylists created new seats that incorporated headrests in an effort to meet another DOT mandate. Lapine's designers also made knobs, handles, sliders, and grips more aggressive. They introduced a new 15-inch-diameter, three-spoke steering wheel with a thick leather wrap.

Shortly after Porsche introduced the G series 1974 models, the world awoke to headlines that had a lasting effect on drivers worldwide. On October 21, 1973, the Organization of Petroleum Exporting Countries (OPEC) announced plans to halt oil shipments to the United States in protest of American political support of Israel. OPEC is a consortium of Middle Eastern oil-producing countries, and the aftereffects of this six-month embargo sent oil and gasoline prices soaring. It had impact at Porsche, where the strength of the sports car market was crucial to its existence.

▲ The 1974 Carrera introduced black window frames and door handles. Rear fenders were slightly wider.

◀ Porsche redid interiors beginning in 1974 models. New seats integrated the headrest into the seat back, and Carreras got this three-spoke steering wheel and standard electric window lifts.

▼ The company introduced "safety stripes" as a front deck lid graphic starting with 1974 Carrera models. The car sold for 37,980DM ($14,721) at the factory, but $13,575 through U.S. dealers.

▲ The second year RS carried over Bosch's mechanical fuel injection, while the base 911 and 911S models got Bosch K-Jetronic. The Typ 911/83 Carrera engine accelerated from 0 to 100 kilometers per hour in 6.3 seconds and reached a top speed of 240 kilometers (149 miles) per hour.

YEAR	**1974-1975**
DESIGNATION	**911**
SPECIFICATIONS	
MODEL AVAILABILITY	Coupe, Targa
WHEELBASE	2271mm/89.4 inches
LENGTH	4291mm/168.9 inches
WIDTH	1610mm/63.4 inches
HEIGHT	1320mm/52.0 inches
WEIGHT	1075kg/2365 pounds
BASE PRICE	$10,475 coupe -$12,233 Targa
TRACK FRONT	1360mm/53.5 inches
TRACK REAR	1343mm/52.9 inches
WHEELS FRONT	5.5Jx15
WHEELS REAR	5.5Jx15
TIRES FRONT	165/70VR15
TIRES REAR	165/70VR15
CONSTRUCTION	Unitized welded steel
SUSPENSION FRONT	Independent, wishbones, MacPherson struts, longitudinal torsion bars, hydraulic double-action shock absorbers
SUSPENSION REAR	Independent, light alloy semi-trailing arms, transverse torsion bars, hydraulic double-action shock absorbers
BRAKES	Ventilated discs, 2-piston cast iron fixed calipers
ENGINE TYPE	Horizontally opposed DOHC six-cylinder Typ 911/92
ENGINE DISPLACEMENT	2687cc/164.4CID
BORE AND STROKE	90x70.4mm/3.54x2.77 inches
HORSEPOWER	150@5700rpm
TORQUE	173lb-ft@3800rpm
COMPRESSION	8.0:1
FUEL DELIVERY	Bosch K-Jetronic fuel injection
FINAL DRIVE AXLE RATIO	4.428:1
TOP SPEED	130mph
PRODUCTION	5,232 coupes: 4,088 Targas all years

1975-1977 TURBO

Porsche's H series 1975 models appeared in October 1974. At the semiannual show in Frankfurt, the production version of the Turbo stopped traffic and started lines behind order takers. A lovely 25th-anniversary coupe commemorating a quarter century of Porsche production got some attention, but Porsche's enthusiasts hung on every specification of the new model: 260 brake horsepower (DIN; roughly 248 SAE net) from a 3.0-liter, 2,993cc (182.6-cubic-inch) engine with 95mm bore and 70.4mm stroke. Porsche quoted acceleration figures from 0 to 100 kilometers in 5.5 seconds, a top speed of more than 250 kilometers (155 miles) per hour, and a cost of DM 65,800 (about $26,750 in 1975).

Performance specifications for base 911, 911S, and Carrera models for European customers did not change. Cars destined for U.S. markets bore new emissions equipment that reduced SAE net horsepower from 167 to 157 for the S and Carrera for 1975. Buyers in California lost another 5 horsepower through thermal reactors, heat exchangers, and an exhaust gas recirculation system, leaving them with 152 SAE net horsepower. Oddly, not every development was bad news for American buyers. A favorable exchange rate for dollars against deutsche marks brought price reductions. The 911S dropped from $12,000 to $11,700, and the Carrera went from $14,000 to $13,600. U.S. cars and those for home markets (except the European Carrera) got electric blowers

◀ Weissach engineers once again bored out the venerable flat six, now to 95mm (3.74 inches), to create 3.0 liters total displacement, 2,994cc (182.6 cubic inches). With the exhaust gas turbocharger, the Typ 930/50 engine developed 260 horsepower at 5,500 rpm.

▲ In addition to adopting the large flat "whale tail" rear wing from the racing RSR models, Porsche also flared front and rear fenders to accommodate rear track widened from base 911 1,340mm (52.8 inches) to 1,511mm (59.5 inches).

▶ The company introduced the Turbo coupe in the spring of 1975. Most cars stayed in Europe, but a number reached Mexico, where this car was delivered.

▲ Porsche offered just two options for the Turbo. One was this graphic, used on the cars the company displayed at Frankfurt and other shows. The other was a sunroof.

for the heater and defroster, improving winter driving comfort in cold climates. Additional insulation retained heat and reduced road noise, an effect abetted by Porsche's decision to run taller final drive gears to improve fuel economy by letting the engine turn slower at comparable road speeds than in 1974 and before. The company offered Targa roof bars in flat black for the first time, as well as the traditional brushed stainless steel.

Porsche exported the Turbo Carrera to the United States as part of the 1976 J series lineup. European buyers got a base 911 with 165 horsepower DIN and a Carrera 3.0 (using the Turbo's 2,993cc engine without the Turbo) with 200 horsepower DIN. For Europe, Porsche dropped the 911S model, while for U.S. customers, the S carried on with all the specifications of the European Carrera model. The U.S. Turbo developed 234 SAE net emissions-controlled horsepower, while the rest of the world still enjoyed 260 DIN. Weissach engineers developed a new four-speed transmission for the Turbo to handle its increased torque and horsepower. Porsche set prices at $25,880 on the East Coast and added $120 in freight to the West Coast, where the Turbo price was $26,000. *Car and Driver* magazine managed to get a test Turbo from 0 to 60 miles per hour in 4.9 seconds, a figure that doubtlessly contributed to first-year sales of 500 in the States (out of 1,300 delivered worldwide).

YEAR	1975-1977
DESIGNATION	**911 Turbo**
SPECIFICATIONS	
MODEL AVAILABILITY	Coupe
WHEELBASE	2272mm/89.4 inches
LENGTH	4291mm/168.9 inches
WIDTH	1775mm/69.9 inches
HEIGHT	1320mm/52.0 inches
WEIGHT	1140kg/2508 pounds
BASE PRICE	$26,750 coupe at factory
TRACK FRONT	1438mm/56.6 inches
TRACK REAR	1511mm/59.5 inches
WHEELS FRONT	8.0Jx15
WHEELS REAR	8.0Jx15
TIRES FRONT	185/70VR15 for 1975, 205/50VR15 for 1976
TIRES REAR	215/60VR15 for 1975, 225/50VR15 for 1976
CONSTRUCTION	Unitized welded steel
SUSPENSION FRONT	Independent, wishbones, MacPherson struts, longitudinal torsion bars, hydraulic double-action shock absorbers
SUSPENSION REAR	Independent, semi-trailing arms, transverse torsion bars, gas-filled double-action shock absorbers
BRAKES	Ventilated discs, 2-piston aluminum front/cast iron rear calipers
ENGINE TYPE	Horizontally-opposed DOHC six cylinder Typ 930/50
	Horizontally opposed DOHC six-cylinder Typ 930/51 (for 1976 US)
ENGINE DISPLACEMENT	2994cc/182.7CID
BORE AND STROKE	95x70.4mm/3.74x2.77 inches
HORSEPOWER	260@5500rpm (930/50)
	245@5500rpm (930/51 w/ airpump, thermal reactors)
TORQUE	253lb-ft@4000rpm (930/50 and 930/51)
COMPRESSION	6.5:1
FUEL DELIVERY	Bosch K-Jetronic fuel injection, turbocharger
FINAL DRIVE AXLE RATIO	4.2:1
TOP SPEED	155mph
PRODUCTION	2,850 coupes (1975 through 1977)

▲ U.S. customers were able to buy the Turbo starting with model year 1976. Dealers charged $25,850.

▲ The Turbo accelerated from 0 to 100 kilometers per hour in 5.5 seconds and was capable of more than 250 kilometers (155 miles) per hour.

▲ The company delivered the cars with Pirelli P7 tires, 205/55VR15 front and 225/50VR15 rear. Optional 16-inch wheels and tires were available.

1976 912 E

For 1976 only, Porsche resurrected the four-cylinder 912, fitted with Bosch's D-Jetronic system, exclusively for the U.S. market. It served as a "placeholder" in the economy sports car world. The company had discontinued its mid-engine 914 in 1975 and didn't introduce its new front-engine, water-cooled 924 until 1977. The injected 912E used the 914's Volkswagen-designed/Porsche-developed 1,971cc (120.2-cubic-inch) flat four developing 86 SAE net horsepower. Porsche priced the car at $10,845 and manufactured just 2,099 as coupes only.

Porsche followed its 25th-anniversary model from 1975 with a 1976 "Ferry Porsche Signature" model, bearing his autograph on the steering wheel. Sales and marketing staffs in Ludwigsburg had learned much from the Carrera RS: Well-heeled buyers throughout Europe and the United States expected a certain level of performance and luxury, and they enjoyed the cachet of exclusivity. Porsche did not manufacture mere automobiles; these were fast "personal" transportation. With its Turbo, the company catered specifically to this customer with an exclusive level of trim and standard equipment. There were, after all, only two options: a sunroof and the Frankfurt show "Turbo" graphic.

For the K series 1977 model year, Porsche introduced chassis improvements that made it easier to steer, brake, and shift gears. Quieter power windows, climate-control air conditioning, automatic speed control, and better road- and wind-noise insulation made long trips and short runs more pleasurable and less fatiguing. European customers had a base 911 available at DM 35,950 for the coupe, DM 38,450

▲ To plug a hole in the entry-level vehicle lineup for the United States, left when Porsche discontinued its 914 before introducing the front-engine 924, the company resurrected the 912, now with fuel injection (*Einspritzung*).

▲ The company manufactured just 2,099 cars, all coupes, which sold for $10,845. The 1,971cc (120.2-cubic-inch) Typ 923/02 opposed four-cylinder engine developed 86 horsepower at 4,900 rpm.

for the Targa. The Carrera 3.0 sold for DM 46,350 in coupe form or DM 48,850 as a Targa. The Turbo, available only as a coupe, was DM 70,000. U.S. buyers chose between the 911S coupe at $14,995 and the Targa at $15,945. The Turbo sold for $28,000.

A Frankfurt-based Porsche mechanic and body-shop owner named Rainer Buchmann capitalized on Porsche's hesitation to offer a Turbo Targa. Starting with a 911SC, his mechanics stripped and reconfigured the car with a 3.3 Turbo engine, 930 flared fenders, and a 1970s-era paint scheme conceived by Eberhard Schulz. Rainer and his brother Dieter's company, b+b Auto-Exclusiv, showed the car at Frankfurt in 1977 and took a number of orders.

In competition, Porsche largely sat out international participation in 1974. It sold Carrera RSR 3.0 models to customers and supported them in European events and the American IROC series. The company had a long history of participation in international rallies, and this continued with the East Africa Safari in 1973 and 1974, with rally veterans Sobieslav Zasada and Björn Waldegaard. Shock absorbers failed (as they had in 1971), and a broken half shaft held Waldegaard to second place overall. But the bulk of Norbert Singer's efforts went to develop turbocharged racing cars, beginning with one prototype 2.1-liter RSR Turbo coupe. (To tame its handling, Singer fitted it with a vast rear wing. Its size embarrassed Fuhrmann, who ordered it painted flat black to make it less obvious.) Porsche had delivered 1,300 Typ 930 production models throughout 1975 and 1976, enough to homologate two RSR variations under FIA regulations.

YEAR	1976
DESIGNATION	912 E
SPECIFICATIONS	
MODEL AVAILABILITY	Coupe
WHEELBASE	2271mm/89.4 inches
LENGTH	4291mm/168.9 inches
WIDTH	1610mm/63.4 inches
HEIGHT	1340mm/52.8 inches
WEIGHT	1160kg/2552 pounds
BASE PRICE	$10,845
TRACK FRONT	1349mm/53.1 inches
TRACK REAR	1330mm/52.4 inches
WHEELS FRONT	5.5Jx15
WHEELS REAR	5.5Jx15
TIRES FRONT	165HR15
TIRES REAR	165HR15
CONSTRUCTION	Unitized welded steel
SUSPENSION FRONT	Independent, wishbones, MacPherson struts, longitudinal torsion bars, hydraulic double-action shock absorbers
SUSPENSION REAR	Independent, light alloy semi-trailing arms, transverse torsion bars, hydraulic double-action shock absorbers
BRAKES	Discs, 2-piston cast iron fixed calipers
ENGINE TYPE	Horizontally opposed OHV four-cylinder Typ 923/02
ENGINE DISPLACEMENT	1971cc/120.3CID
BORE AND STROKE	94x71mm/3.70x2.80 inches
HORSEPOWER	86@4900rpm
TORQUE	98lb-ft@4000rpm
COMPRESSION	7.6:1
FUEL DELIVERY	Bosch L-Jetronic fuel injection
FINAL DRIVE AXLE RATIO	4.428:1
TOP SPEED	109mph
PRODUCTION	2,099 coupes

▲ (Top) Porsche sold these race cars for 108,000DM, roughly $41,300 at the time. The racing department assembled 30 of the 934s in 1976.

▲ (Bottom) Air-to-water intercoolers in the front air dam cooled the fuel-air mixture enough to develop 485 horsepower out of the 2,993cc (182.6-cubic-inch) Typ 930/25 engine at 7,000 rpm. Later in the 1976 season, these engines produced 580 horsepower.

▲ This was the strictly business 911 Turbo for the racetrack: the Group 4 Typ 930. Wheel extensions added 50mm (1.97 inches) to front and rear track. The front air dam ventilated brakes, two intercooler radiators and an oil cooler in the center.

1976–1977 934 AND 935

This was a time of cross-pollination in series production and race car development. Wolfgang Berger worked with Norbert Singer to develop the "silhouette" 934 and 935 models, 930 Turbo–based race cars for FIA groups 4 and 5. Singer needed taller and wider wheels and tires on these cars, which brought 16-inch rubber to the street 930 for model year 1977. Singer's engineers created the Typ 934, the Group 4 930.

▲ 1977 911S Coupe
This was the final year for the legendary 911S model. U.S. cars suffered further emission controls with secondary air injection pumps, thermal reactors, and the first-generation catalytic converter.

▶ Interiors received a few upgrades. New air vents appeared in the center of the dash. Door lock buttons disappeared into the doorsills when locked to thwart attempts to open cars with wire coat hangers.

This uncompromised race car competed with electric window lifts, because Porsche's series production models sold to enthusiastic European and American customers with these as an expected luxury item. Singer maintained that the electric lift mechanism weighed less than a manual crank and gears. The 934 sold for DM 97,000 ($40,000), complete with a 3.0-liter engine with K-Jetronic induction that developed 485 DIN horsepower. To wrench this much additional horsepower from the 930 engine, Porsche introduced air-to-water intercoolers, one for each bank of cylinders. (The 1974 show prototype had one intercooler only.) The 934 weighed 1,207 kilograms (2,655 pounds), compared to the street 930 at 1,266 kilograms, or 2,785 pounds. Porsche bolted on large fender flares to accommodate 23.5 x 10.5-16 front and 25.55 x 12.5-16 rear racing tires. A pair of 934s took first and second in the season points championship for the Sports Car Club of America's Trans-Am Challenge.

Porsche's Group 5 "special production car" entry, the Typ 935, was more extreme. Its 2.85-liter (173.9-cubic-inch) engine produced 590 horsepower within a 1,064-kilogram (2,340-pound) race car that Porsche sold for $75,000. Fiberglass panels replaced steel everywhere possible, except in the silhouette-holding roof. The single air-to-air intercooler filled a massive rear wing

▲ Conceived as an affront to the under-2.0-liter racing class, Porsche developed an ultra-light (and slightly shrunk) racer nicknamed "Baby." Its 1,425cc (86.9-cubic-inch) turbo motor developed 370 horsepower in a 750-kilogram (1,650-pound) car. In its second race, it won by 52 seconds, and Porsche retired the car, having made its point. *Photograph courtesy Porsche Archive*

that satisfied Fuhrmann only when FIA regulators said its width had to fit within overall car dimensions. Creative reading of the other regulations led Singer and Berger to flatten the front fenders so that they barely cleared the front tires. They installed headlights in the low front spoiler and cut louvers over the tires to vent air pressure.

Road & Track magazine ran side-by-side tests, accelerating a 934 and a 935 from 0 to 60 and on to 100 miles per hour. The 934 reached 60 in 5.8 seconds. The 935 hit 60 in 3.3 seconds and 100 in 6.1—4 seconds ahead of the 934.

When Ernst Fuhrmann got to Porsche, the company's products had invisible targets on their backsides. Worldwide regulations for safety and exhaust emissions threw into question the viability of rear-engine air-cooled sports cars in the future. Fuhrmann hedged his bets and shifted Porsche's development efforts toward front-mounted engines with water cooling. The entry-level 924 reached markets in 1976. The 928, the company's flagship, arrived as a 1978 model. Both of these looked, sounded, and performed radically differently from the 911. In Fuhrmann's mind, these represented what the company products needed to be. He and the supervisory board put another mark on a calendar designating the end of the 911.

Throughout 1977 and early 1978, Fuhrmann fielded questions about the company's product lineup. Porsche was assembling about 45 911s each day. To quiet his inquisitors, he explained his plan. He clarified it again in an interview in 1991.

▲ The Jägermeister-sponsored 934, entered by Herve Poulain and prepared by the Kremer Brothers, was one of 55 starters at the June 1978 Le Mans 24 Hours Race.

▶ The second-generation Typ 930/73 engines developed nearly 600 horsepower at 7,200 rpm, due to many internal changes. Engineers increased boost from 1.3 bar (18.5 psi) to 1.7 bar (24 psi).

▲ Porsche adopted the air-to-water intercoolers used on 934s for this more aggressive Group 5 car. Its 2,856cc (174.2-cubic-inch) turbocharged Typ 930/72 engine developed 590 horsepower at 7,900 rpm at 1.5 bar boost, 22 psi.

▲ In Cologne, brothers Erwin and Manfred Kremer had begun racing Porsches in 1971. By the time the 935 appeared in 1975, they were well established as drivers, car builders, and preparers.

"The car was still selling. We still made money from this car," he said. "So I set a low limit at which we no longer make money. I told journalists if we ever go below twenty-five cars . . . six thousand a year, we stop. That quieted them."

He didn't explain that he already had suspended further development on 911s beyond meeting ongoing U.S. emission and safety regulations. America still absorbed half of 911 sales, so half its profit came from satisfying U.S. demand. The 924 brought in a younger audience that never believed it could own or drive a Porsche. The 928 reintroduced Porsche as a luxury carmaker. For 1978 the 911 G series introduced the Super Carrera, the SC. Few people outside Porsche knew that this model was born with an invisible target on its back.

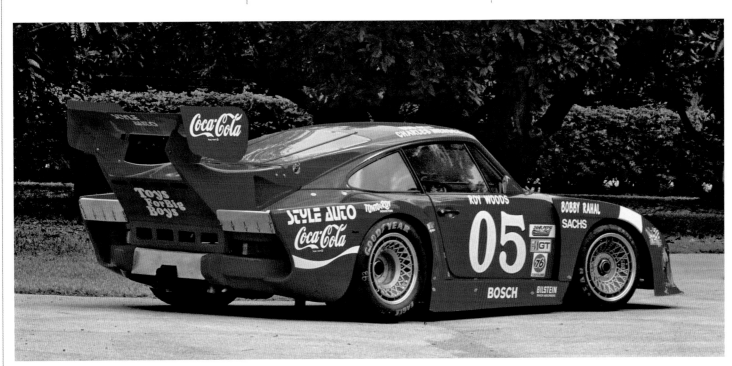

▲ Beginning with Porsche's 1977 935/77 customer race cars, the Kremers made hundreds of small and large changes to produce dominating racers. By 1980, their engines, bored to 3.3 liters, 201 cubic inches, developed 800 horsepower at 8,000 rpm.

▲ While the racing 934 carried a high minimum weight (1,120 kilograms—2,464 pounds), engineers simplified the instrument panel to just oil temperature and pressure, engine speed, and turbocharger boost.

▲ (Top) With Bob Akin, Roy Woods, and Bobby Rahal at the wheel, the Hudson Wire–sponsored 935 started 19th on the grid for the 1980 Sebring 12 hours event and finished 5th.

▲ (Top) Looking very much like a Porsche Kremer 935 racer, this road car came by its appearance honestly. Ekkehard Zimmerman, founder of dp designed the car bodies for Kremer racing. *Photograph © 2011 Steve Mraovic*

▲ (Bottom) The car was manufactured for a German buyer in 1977, and in the early 1980s, O'Gara Coachworks of Beverly Hills, California, imported it and sold it to rock singer Rod Stewart, who enjoyed it for many years. *Photograph © 2011 Steve Mraovic*

YEAR	**1976-1977**
DESIGNATION	**934 and 935**
SPECIFICATIONS	
MODEL AVAILABILITY	Coupe
WHEELBASE	2268mm/89.3 inches
LENGTH	4235mm/166.7 inches
WIDTH	1775mm/69.9 inches
HEIGHT	1320mm/52.0 inches
WEIGHT	1120kg/2470 pounds
BASE PRICE	$41,300 coupe at factory for 1976
TRACK FRONT	1481mm/58.3 inches
TRACK REAR	1496mm/58.9 inches
WHEELS FRONT	10.5Jx16
WHEELS REAR	12.5Jx16 (up to 17.0Jx16 for 1979/1980)
TIRES FRONT	275/600-16
TIRES REAR	325/625-16
CONSTRUCTION	Unitized welded steel
SUSPENSION FRONT	Independent, wishbones, MacPherson struts, longitudinal torsion bars, hydraulic double-action shock absorbers
SUSPENSION REAR	Independent, semi-trailing arms, transverse torsion bars, gas-filled double-action shock absorbers
BRAKES	Ventilated discs from Typ 917, 4-piston aluminum calipers
ENGINE TYPE	Horizontally opposed DOHC six-cylinder Typ 930/25
ENGINE DISPLACEMENT	2994cc/182.7CID
BORE AND STROKE	95x70.4mm/3.74x2.77 inches
HORSEPOWER	485@7000rpm; 580@7000 for 1977
TORQUE	475lb-ft@5400rpm
COMPRESSION	6.5:1
FUEL DELIVERY	Bosch K-Jetronic fuel injection, turbocharger, intercooler
FINAL DRIVE AXLE RATIO	various
TOP SPEED	190mph
PRODUCTION	32 coupes in 1976; 10 coupes in 1977

▲ With close to 500 horsepower on tap, the car had the performance to match its looks. Each of the *dp* 935s was commissioned and personalized for its initial buyer. *Photograph © 2011 Steve Mraovic*

1978–1979 **911SC**
1978 **935-78 "MOBY DICK"**
1978–1979 **911 TURBO**
1980–1983 **911SC AND** 1980–1985 **911 TURBO**

CHAPTER 3

THE SECOND GENERATION 1978–1983

1978-1979 911SC

Five years after management planned to kill it, Porsche introduced the next *final* 911. The company commingled nomenclature from the discontinued—and lamented—benchmark S with the more recent benchmark-setting Carreras. The result was a hybrid designation, the SC. Whatever the abbreviation meant to the marketing staff in Ludwigsburg—Super Carrera, Special Carrera, or Sport Carrera—it signified to 911 enthusiasts that Porsche had saved the car. It was still here, and it was better.

The company put its efforts into the existing 2,993cc flat six introduced in the 1976 Carrera. Weissach dedicated itself to increasing torque. While horsepower ratings from the earlier Carrera dropped (from 210 DIN to 180, or 172 SAE net, due to mandatory emissions-controlling air pumps), the torque rose above either of the car's predecessors, going from 188 DIN to 195 DIN at 4,200 rpm (and 189 lb-ft SAE net for U.S. customers). Engineers stretched the useful range with a new crankshaft and larger rod and main bearings. They adopted a breakerless ignition and programmed it to run smoothly on regular unleaded gasoline. The same Bosch system that controlled spark also guided a more accurate and effective rev limiter. Porsche abandoned its 5-blade cooling fan and returned to the more efficient (and quieter) 11-blade version. The company offered the coupe and Targa with a five-speed manual transmission or the three-speed Sportomatic. The 930 Turbo was available only with the four-speed gearbox.

As one means of launching the new model, Porsche revisited the East

◀ All U.S. 1978 SC models ran Typ 930/04 engines with an air pump (at far left in the engine compartment) to meet emissions standards. All cars for all markets developed 180 horsepower at 5,500 rpm, good to get the cars from 0 to 100 kilometers per hour in 7.0 seconds.

Africa Safari. Roland Kussmaul entered two heavily reinforced 911SCs for 1978. The event covered 3,000 miles (4,800 kilometers), little of it paved, through deserts and rain forests. The preparation paid off, and despite brutal terrain that battered and broke suspensions, Porsche finished second and fourth.

▶ Porsche entered two cars in the 26th Safari Rally that started and ended in Kenya. Lightened but heavily reinforced, the cars weighed 1,180 kilograms, 2,596 pounds, with all equipment, and were geared to run as fast as 228 kilometers (142 miles) per hour.

◀ Porsche gave the car nearly 280mm (11 inches) of ground clearance. Even with shielding, suspension arms were vulnerable, and a rear arm failed on this car while Preston's car suffered delays repairing broken half shafts. Waldegård finished fourth.

▲ Porsche fitted modified production Typ 911/77 engines, prepared and tuned to develop 250 horsepower at 6,800 rpm. Previous Safari winner (and Monte Carlo winner) Björn Waldegård had driving duties, while Vic Preston drove the other car.

▲ To create the new SC, Porsche combined the best features of the earlier S and Carrera models, including the Carrera's wider flares. The Targa sold for 42,700DM at the factory and $20,775 at U.S. dealers.

◀ Sometime during the 1978 model year production run, the company switched over to 1979-style body color headlight bezels and black-painted Targa bars. The interiors were largely unchanged.

YEAR	**1978-1979**
DESIGNATION	**911SC**
SPECIFICATIONS	
MODEL AVAILABILITY	Africa Safari
WHEELBASE	2272mm/89.4 inches
LENGTH	4291mm/168.9 inches
WIDTH	1652mm/65.0 inches
HEIGHT	1450mm/57.1 inches
WEIGHT	1180kg/2601 pounds
BASE PRICE	Not available
TRACK FRONT	1369mm/53.9 inches
TRACK REAR	1379mm/54.3 inches
WHEELS FRONT	9.0Jx15
WHEELS REAR	11.0Jx15
TIRES FRONT	215/60VR15
TIRES REAR	235/60VR15
CONSTRUCTION	Unitized welded steel
SUSPENSION FRONT	Independent, wishbones, MacPherson struts, longitudinal torsion bars, hydraulic double-action shock absorbers
SUSPENSION REAR	Independent, light alloy semi-trailing arms, transverse torsion bars, hydraulic double-action shock absorbers
BRAKES	Ventilated discs, 2-piston cast iron fixed calipers
ENGINE TYPE	Horizontally opposed DOHC six-cylinder Typ 911/77
ENGINE DISPLACEMENT	2994cc/183.2CID
BORE AND STROKE	95x70.4mm/3.74x2.77 inches
HORSEPOWER	250@6800rpm
TORQUE	217lb-ft@5500rpm
COMPRESSION	9.1:1
FUEL DELIVERY	Bosch K-Jetronic fuel injection
FINAL DRIVE AXLE RATIO	3.875:1
TOP SPEED	142.5mph
PRODUCTION	4

1978 935-78 "MOBY DICK"

While racing in Africa was grueling, the cars were not extraordinarily different from series production models. However, they were completely unlike another competition vehicle that emerged from Porsche's racing shops for 1978. Norbert Singer, the racing engineer with a talent for reading FIA competition rule books, designated the car as the 935/78. But when he first rolled it out for racing journalists to photograph late in the winter, it earned the nickname Moby Dick.

It was radical. There was a resemblance to the 911, but this was longer, lower, wider, and much wilder than anything out of the racing—or the styling—department yet. It came to life because Singer read not only what the FIA had written but what it had forgotten to include.

Helmuth Bott had his doubts. "Are you sure you can do this?" he asked.

"Well, by the letter, I am quite sure," Singer replied.

FIA "letters" dictated that "the original body shape must be retained, doors, and roof." It went on, "Fenders are free," meaning constructors could redesign them as they desired. That had been the loophole that allowed Singer to flatten the 911's characteristic stovepipe fenders into louvered slabs for the 935, leaving the race car's headlights inches above the track. These newest regulations invited Singer to widen fenders to accommodate fatter tires, to add a second door skin flush with the broader fender lines, and to fit a new, flatter fiberglass roof above the original steel panel that directed airflow smoothly to the longest tail yet seen on a 911. Singer sliced off the doorsills, dropping the car 6 centimeters (2.4 inches) closer to the ground. He substituted the floor pan with a thin layer of fiberglass attached to a new aluminum tube frame.

While Singer stretched the bodywork in all directions, Hans Mezger and his colleagues wanted to incorporate four valves in the heads, but this left too little room for cooling fins. They concluded that the best way to cool the engine was with water. They devised a system on a 2.8-liter motor where water entered each four-valve head from the bottom and a pump forced it out the top. Manufacturers had to use "stock" cylinder blocks, though, as Jürgen Barth understated it, rules "allowed some additional machining."

Experiments and testing proved the concept a success, so Weissach developed 3.2-liter (195.1-cubic-inch) blocks and a 2.1-liter (128.1-cubic-inch) engine for a new 936 sports racer for Le Mans. (At the same time, Porsche pursued an Indianapolis 500 effort for 1979 and 1980 with American Ted Fields and his Interscope Racing team. Engineers developed several 2.65-liter/161.7-cubic-inch, alcohol-fueled flat sixes with water-cooled heads for that contest.) If that were not enough challenge, Mezger's crew also developed a 1.4-liter (85.4-cubic-inch) flat six to install in a scaled-down 935 nicknamed Baby to take on smaller-displacement BMWs and other competitors. As Barth explained, "These water-cooled and air-cooled engines were almost completely new. From the original Nine-Eleven, engineers took only a modified crankshaft with larger bearings. Pistons, connecting rods, gear-driven camshafts, were all new."

Moby Dick debuted at the season-opening race in Italy, a country with a long memory for moments when Enzo Ferrari's loyal fans believed he had been cheated against. The protest five years earlier that allowed Singer to upgrade Porsche's production-based RSR into the prototype category and win a season led him to move pre-emptively. The FIA routinely offered to preview race cars. One member of the

▲ As raced, the car used a third variety of rear wing. Martini & Rossi sponsored factory-racing 935s for many years.

committee who arrived to inspect the 935/78 was longtime Porsche racer and friend Paul Frère.

"Illegal," he said. "The rule book says you have to have the production shape of the door, not these funny. . ."

"No problem," said Singer, undoing the quick-release fittings. "Underneath is the original door." His staff then removed all the extra panels from the car.

"The shape was there," he explained years later, "and there was no sentence in the rule book about whether you could cover it." Frère and his two colleagues stewed and then concluded that because what Singer had done was not specifically illegal, it was legal. They insisted, however, that Porsche run the car with the original doors exposed. Singer insisted that the inspectors approve the car in writing before they left, signing the forms with photos attached.

Mezger's 935/71 engines developed 750 DIN horsepower at 8,200 rpm from 3,211cc displacement. The entire car weighed 1,025 kilograms, or 2,255 pounds. Racing at Le Mans in June, in striking Martini & Rossi racing livery, the car regularly hit 365 kilometers per hour (228 miles per hour) on Mulsanne. Jürgen Barth, one of the factory team drivers and head of Porsche's customer racing program, was first to drive it.

▲ FIA inspectors, invited to Porsche to preview this car, greeted it with howls: "It can't be! It's illegal." Somewhere under all the wider bodywork was an actual 935. *Photograph courtesy Porsche Archive*

▲ Regulations said the roof could not be changed. But after Norbert Singer added wider door panels and an airplane-size wing, FIA regulators made him return to actual doors and reduce the wing. *Photograph courtesy Porsche Archive*

▶ When the rules mentioned nothing about doorsills, Norbert Singer sliced them off, lowering the car 60mm (2.4 inches). He sacrificed the contoured floor pan for an aluminum tube frame covered with a thin piece of fiberglass.

▲ Cologne-based real-estate developer Georg Loos hired Porsche mechanics to work his team on race weekends. It served him well, winning the 1000 kilometers of Nürburgring in May 1978, shown here, among many other events. *Photograph courtesy Porsche Archive*

▲ Water-cooling allowed engineers to create four-valve cylinder heads on this 3,211cc (195.9-cubic-inch) Typ 935/3.2 flat six. Huge twin turbochargers run at 1.8 bar (26 psi) boost delivered 750 horsepower at 8,200 rpm.

"This was the first race car I ever drove," he recalled in an interview in 2004, "that I held onto the wheel with both hands. With boost all the way up, I think it was more than nine hundred horsepower. I think one hundred sixty kilometers [100 miles per hour] came in six seconds, maybe less. In second gear! Then shift to third and fourth and it starts all over again each time."

But Moby Dick failed to meet expectations throughout the 1978 season. It didn't finish its first race, and it didn't compete at Norisring in Germany or at Vallelunga in Italy later in the year. It placed only eighth at Le Mans. Finally, at Silverstone in England, it won easily. Porsche produced only two cars, plus the tubular space frame for a third. It was several years before copies appeared and did the concept justice. Privateers John Fitzpatrick from England and Gianpiero Moretti from Italy created and raced their own long-tailed Moby Dicks with help from outsiders such as Erwin and Manfred Kremer from Cologne. For Barth's customers, Porsche produced a more restrained-looking 935/78, and these cars often outshined the more dramatic Moby Dick.

YEAR	1978
DESIGNATION	**935-78 Moby Dick**
SPECIFICATIONS	
MODEL AVAILABILITY	coupe
WHEELBASE	2279mm/89.7 inches
LENGTH	4890mm/192.5 inches
WIDTH	1970mm/77.6 inches
HEIGHT	1265mm/49.8 inches
WEIGHT	1030kg/2266 pounds
BASE PRICE	Not available
TRACK FRONT	1630mm/64.2 inches
TRACK REAR	1575mm/62.0 inches
WHEELS FRONT	11x16
WHEELS REAR	15X19
TIRES FRONT	275/600-16
TIRES REAR	350/700-19
CONSTRUCTION	Aluminum tube frame
SUSPENSION FRONT	Independent, wishbones, coil-over gas-filled shock absorbers, adjustable anti-roll bar
SIISPENSION REAR	Independent, light alloy semi-trailing arms, titanium coil springs, gas-filled double-action shock absorbers, adjustable anti-roll bar
BRAKES	Ventilated, drilled discs from Typ 917, 4-piston aluminum fixed calipers
ENGINE TYPE	Horizontally opposed DOHC six-cylinder Typ 935/3.2
	Water-cooled cylinder heads, turbocharged, intercooled
ENGINE DISPLACEMENT	3211cc/195.9CID
BORE AND STROKE	95.7x74.4mm/3.77x2.93 inches
HORSEPOWER	650@8200rpm
TORQUE	615lb-ft@6500rpm
COMPRESSION	7.0:1
FUEL DELIVERY	Bosch mechanical fuel injection
FINAL DRIVE AXLE RATIO	Varied by circuit
TOP SPEED	220mph
PRODUCTION	2

1978-1979 911 TURBO

As early as 1971, Erwin Kremer and his brother Manfred had prepared, raced, and won in Porsches from their Auto Kremer shops in Cologne. In his 911S, Erwin claimed the Porsche cup that year. A Georg Loos/Erwin Kremer–prepared RSR took John Fitzpatrick to the European GT championship in 1974. The Kremer brothers responded when Porsche offered 934s to outside customers, and with Bob Wollek at the wheel, their 934K dominated the European GT season in 1976. The car's radical styling and successful aerodynamics were other elements of its success, a creation of Cologne designer Ekkehard Zimmerman and his company Design + Plastic. When customer 935s appeared in June 1978, the brothers went to work. They debuted their 1979 935K3 at Zolder in Holland. For weight reduction and cooling efficiency, they replaced Porsche's water-to-air intercooler with an air-to-air version. With other internal changes, they claimed 805 brake horsepower at 8,000 rpm.

To increase stiffness, they welded nearly 100 feet of aluminum tubing onto the internal roll cage and added more out front to extend Zimmerman's bodywork and in back to provide better engine access for repair or removal. They reskinned the car in DuPont

▲ The Typ 930/60 engine accelerated the 1,300-kilogram (2,860-pound) coupe from 0 to 100 kilometers per hour in 5.4 seconds. Top speed was 260 kilometers (161 miles) per hour.

▲ Little changed on the Turbo body, but underneath that rear wing, a new engine displaced 3,299cc, 201.2 cubic inches. With a new charge-air intercooler just beneath the louvers, the Typ 930/60 engine developed 300 horsepower, while U.S. Typ 930/61 engines developed 265 horsepower at 5,500 rpm.

Kevlar, which lightened the 935K3 by 30 kilograms (66 pounds). But, as Karl Ludvigsen reported, this increased the cost of the body by ten times. As Manfred Kremer told Paul Frère in *Road & Track*, they made about 100 changes in their K3 compared to the factory-delivered 935, which "add up to one percent more efficient car, which is all you need to beat the competition." For each race, they used a freshly rebuilt engine and updated details throughout their cars. Kremer team driver Klaus Ludwig won Le Mans in 1979, and other K3s came in third and eleventh. Ludwig won 11 of 12 races to seize the European GT championship. With a record like that, customers lined up to buy K3s at DM 375,000, roughly $210,000 to U.S. competitors. Series production models for the 1978 L series and the 1979 M series remained reserved and contained by comparison to how the racers evolved. Factions sprung up within styling and engineering departments. Those who loved the 911 were frustrated that management had halted engineering development and design updates. Wolfgang Möbius, who designed the 928, still drove a 911 as his chosen company car. Production of 911SC and Turbo models for the L series reached 10,743 units; this climbed to 11,596 cars for 1979, though the models barely changed.

▲ Short-pile carpeting replaced first-generation thicker material. The Turbo got tinted windows all around. The car sold for 79,900DM and $36,700 in the United States.

THE SECOND GENERATION 1978–1983

YEAR	1978-1979
DESIGNATION	**911 Turbo**

SPECIFICATIONS

MODEL AVAILABILITY	Coupe
WHEELBASE	2272mm/89.4 inches
LENGTH	4291mm/168.9 inches
WIDTH	1775mm/69.9 inches
HEIGHT	1310mm/52.4 inches
WEIGHT	1300kg/2860 pounds
BASE PRICE	$43,661
TRACK FRONT	1432mm/56.4 inches
TRACK REAR	1501mm/59.1 inches
WHEELS FRONT	7.0Jx16
WHEELS REAR	8.0Jx16
TIRES FRONT	205/55VR16
TIRES REAR	225/50VR16
CONSTRUCTION	Unitized welded steel
SUSPENSION FRONT	Independent, wishbones, MacPherson struts, longitudinal torsion bars, gas-filled double-action shock absorbers, anti-roll bar
SUSPENSION REAR	Independent, light alloy semi-trailing arms, transverse torsion bars, gas-filled double-action shock absorbers, anti-roll bar
BRAKES	Ventilated, drilled discs, 4-piston aluminum calipers
ENGINE TYPE	Horizontally opposed DOHC six-cylinder Typ 930/60
	Horizontally opposed DOHC six-cylinder Typ 930/61 (1977/1978 US 49-states w/ thermal reactors, EGR)
	Horizontally opposed DOHC six-cylinder Typ 930/63 (1977/1978 US California w/ thermal reactors, EGR, vacuum ignition retard)
ENGINE DISPLACEMENT	3299cc/201.3CID
BORE AND STROKE	97x74.4mm/3.82x2.93 inches
HORSEPOWER	300@5500rpm (930/60)
	265@5500rpm (930/61 and 930/63)
TORQUE	304lb-ft@4000rpm (930/60)
	291lb-ft@4000rpm (930/61 and 930/63)
COMPRESSION	7.0:1
FUEL DELIVERY	Bosch K-Jetronic fuel injection, turbocharger, intercooler
FINAL DRIVE AXLE RATIO	4.22
TOP SPEED	160mph
PRODUCTION	14,476 worldwide 1978 – 1988. Turbo not available legally in US from 1980 through 1985

▲ (Bottom) **1983 930 Turbo M505** Porsche still did not export its 930 Turbo models to the United States or rest-of-the-world customers, but the company offered the M701 option, the flat nose, or *flachtbau*, to those who could buy the cars. Initially it was available on new orders only for an additional 38,340DM, about $15,032 at the time.

▲ (Top) Side sills and rear fender air inlets added an additional 10,772DM, $4,224, to the price. A fully optioned Turbo with bodywork and a "performance kit" that boosted horsepower to 330 at 5,750 rpm cost 183,272DM, $71,871 at the time.

1980-1983 911SC AND 1980-1985 911 TURBO

To commemorate the opening of Porsche's Weissach engineering and design center and test track, the company issued a 180-horsepower (SAE) "911 Weissach Edition" for U.S. customers for 1980 as part of the new-designation A Program. Porsche fitted the cars, available either in platinum metallic with platinum wheels or in black metallic, with special interior appointments and the Turbo whale tail. It limited output to 408 cars.

A slightly more special car, the 911 SC 3.1, appeared at the same time, limited to about 100 copies. Porsche engineers fitted it with engines bored to 3.3-liter Turbo dimensions—97-mm—while retaining 3.0-liter SC engine stroke at 70.4mm. This yielded 3,122cc, or 190.4 cubic inches, and the engine developed 210 DIN horsepower

▲ With the 1980 model, the 2,994cc (182.6-cubic-inch) electronically fuel-injected Typ 930/09 engine developed 188 horsepower at 5,500 rpm, an increase of 8 horsepower. The car sold for $32,000 in the Untied States, with 180 horsepower.

at 5,800 rpm. Some 50 of these cars remained in Porsche engineering, and there is some evidence that the engines were development mules for the chain tensioner system Porsche introduced on the 3.2 Carreras in 1984. While all the cars carried over Bosch K-Jetronic mechanical fuel injection, a number of the 3.1s appeared with dual exhausts, and many of them bore Turbo front and rear spoilers, similar to the U.S-destined Weissach Editions. Porsche apparently sold the other half of 3.1 production to favored customers.

Engineers dealt with emission and safety standards. These included two separate specifications for U.S. deliveries, one set for California cars and a second for the other 49 states. The EPA rationalized the two for 1980, adopting *lambda sonde* oxygen sensors and three-way catalytic converters. But emissions restrictions still constrained U.S.-bound cars to 180 SAE net horsepower. Porsche management faced hard choices as it struggled to meet worldwide regulations. For the U.S. markets, saddled with a 55-mile-per-hour national speed limit, safety consciousness and fuel consumption concerns forced Porsche to fit speedometers that read only to 85 miles per hour. The company withdrew the 930 Turbo from U.S., Canadian, and Japanese markets. The three-speed Sportomatic transmission disappeared as well, a victim of changing tastes and diminishing desires. Stuttgart factory output fell to 9,943 cars for 1980, leaving some distance to go before the model fell against Ernst Fuhrmann's arbitrary 6,000-unit death sentence. But Fuhrmann split advertising, engineering, and design budgets three ways. While the 911 and 930 constituted a third of total production (including 924 and 928 models) of 28,622 cars, observers recognized that the air-cooled cars got less attention.

In his last years at Porsche, Ernst Fuhrmann made few friends and earned more enemies. In an interview in 1991, he admitted that he had caused some of that himself. His reputation fell further as people criticized him for usurping the role of company spokesperson from Ferry Porsche at public events. Yet this was an element of the job description, and Fuhrmann's successors have filled it as well. Senior staff inside the company who knew Ferry Porsche well said that he was soft-spoken, introverted, and circumspect. He never enjoyed being the center of attention.

Ferry's strengths were beyond dispute. Whether or not his father ever intended to produce a Porsche automobile, Ferry did, and he saw it through birth and success. He managed a difficult family meeting in 1970 that removed everyone in his growing family from roles in the company, except for himself. Those were not the accomplishments of a weak man. However, for reasons he never explained, he exiled himself from his own company.

Early in 1979, Ernst Fuhrmann anticipated his 60th birthday. He looked back on his two careers at Porsche,

▲ Along with the rear whale tail, Weissach coupes came with sport shock absorbers, an electric sunroof, and electrically heated outside mirrors.

▲ This interior combination of Doric Gray leather with burgundy piping was unique to the Weissach SC. For U.S. customers, Porsche fitted a speedometer reading only to a maximum of 85 miles per hour.

▶ Porsche assembled 400 of these primarily for U.S. sales. The car was available only in this Platinum Metallic or Black, in equal numbers.

▲ Dick Barbour and John Fitzpatrick became a potent force in IMSA and World Endurance Championship races starting in 1980 with this third iteration of the Kremer Brothers' 935 creations.

as engine designer and as company leader. The front engined, water-cooled 924 that he had conceived was three years along and generating more than 20,000 sales each year. His flagship 928, then two years old, had earned respect as a potent grand tourer. Its production approached half the output of the 911, which soldiered on, approaching its 15th birthday as he reached number 60. But he knew company history: On the 15th anniversary of the 356, the company had shown the world its successor.

"The new cars," Fuhrmann recalled in 1991, "the Nine-Twenty-Four, Nine Twenty-Eight, they were through the program. Work was a little slow. At that time we should have begun a new program." He told Ferry Porsche that he wanted to retire at 65, and since a new development sequence took seven or eight years, he did not want to begin something and leave it uncompleted for his successor. "I'm prepared to go," he told his boss, "any day, if you have a new man who could begin again with a new program."

But Porsche had little to offer anyone new. Poor morale colored its union workers' attitudes. At the opposite corner of Stuttgart, Mercedes-Benz had taken on Porsche's 928 with its new 2+2 450SLC. The two-seat 450SL coupes and roadsters were 140-mile-per-hour autobahn cruisers that chased the 928s and 930 Turbos and challenged 911s. M-B production exceeded 400,000 cars. From Munich, BMW introduced a road-going version of its M1 racer for loyal customers who sought something faster and sleeker than its 2+2 633CSi coupes. BMW assembled 330,000 cars in 1980.

Ferry Porsche contacted Bob Lutz about the job. Lutz had run operations at Ford of Europe in Cologne for several

◀ The Porsche-built/Kremer modified 3,164cc (193.0-cubic-inch) twin turbocharged engine developed as much as 800 horsepower when drivers increased the boost to 1.8 bar (26 psi). K3s reached 385 kilometers (240 miles) per hour along the Mulsanne Straight at Le Mans.

▼ In this configuration with Brian Redman as co-driver, this car finished fifth overall, first in Le Mans GTP class, in 1980. John Fitzpatrick then finished second in the Endurance Driver's Championship, and Dick Barbour was third.

years. He sometimes drove a Turbo, but his conversations with Porsche gave him little understanding of where Ferry thought the company should go. No new projects meant no new cars for at least five or six years. Could Porsche keep the 911 going till 1985 or 1986? Lutz passed. Ferry continued looking. One after another candidate paused, looked, and walked on. Finally, after satisfying his doubts, one candidate, a German engineer educated in the United States, accepted the invitation to meet the Porsche and Piëch families. It took very little time for them to determine whether the new man had the courage and imagination to walk into Fuhrmann's footsteps—or to break a new trail all his own.

Peter Schutz had left an engineering job at Caterpillar because he had ideas to help the company make money, but management wanted him developing engines.

Schutz had never owned a Porsche. But he did his homework before he started at Zuffenhausen. Talking to dealers across America and Europe, he learned their two complaints: Porsches were too expensive, and they had quality-control problems. To Schutz, this was a single issue: No buyer objected to paying any price for something, so long as it functioned perfectly. He discovered that morale at Porsche had plummeted because it was discontinuing the 911 to push the 924 and 928. He sensed there was more. In January 1981, he and his wife, Sheila, arrived in Zuffenhausen. He immersed himself in information, discovery, and action.

Schutz believed that taking the car out of production was what struck him as wrongheaded. Finance figures showed that the 911 was the company's most profitable car. Even so, Porsche had no plans to build the 911 after 1981.

▲ The U.S. Department of Transportation forced auto manufacturers to change speedometers to read a maximum of 85 miles per hour. Gasoline shortages and price increases encouraged conservation through a 55-miles-per-hour national speed limit.

▼ The M473 option, "With Spoilers," installed the Turbo "tea tray"–type rear wing on the SC coupes for the first time on 1982 models. It was popular in the UnitedStates, where buyers could not buy a Turbo through authorized dealers.

▲ For those moments when cars posed for pictures, the new standard 1,050-watt alternator with internal voltage regulator was a welcome improvement. Porsche added it to cope with electric windows, sunroofs, air conditioners, and aftermarket stereo and theft alarm installations.

▲ Most significantly for 1981 model year, Porsche and Bosch revised the K-Jetronic fuel injection system to incorporate a cold-start injector spray that prevented backfires that destroyed airboxes, stalling the car in place. The new 930/10 engines developed 204 horsepower for the rest of the world.

▲ (Top) SC coupes for 1982 sold for 51,850DM, roughly $21,338, at the factory. Acceleration from 0 to 100 kilometers per hour took 6.8 seconds, and Porsche quoted the top speed at 235 kilometers, 146 miles per hour.

▲ For the first year, 1983, Porsche offered the convertible top only in black. Leather front seats were standard for the Cabriolet.

Thus the transition from 1979 to 1980 to 1981 had brought few changes. Fuhrmann had allowed repairs but no replacements. Under Schutz, 1981 C Program cars got spring-steel clutches to replace failure-prone (but gentler operating) rubber-centered ones. Backfires on start-up often destroyed airboxes, immobilizing cars, so Porsche and Bosch reprogrammed the K-Jetronic injection cold-start mixture.

It had been 18 years since Porsche introduced its 901 at the biennial auto show in Frankfurt, and the company's image had slipped. It was time for Porsche to ignite public imagination.

Bott and Schutz hoped to introduce the 911 cabriolet as a 1982 model. Stiffening the chassis and finessing the top mechanism took time. The car appeared for 1983, the final year of the SC series, the last of the second-generation 911.

Gerhard Schröder had designed the top mechanism for F. A. Porsche's concepts in 1963. He tackled this new assignment: Ferry Porsche wanted the top to operate electrically. Engineering had a working prototype by March 1982, but it needed more work. For the 1983 911SC cabriolet, owners opened or closed the top by hand and zipped a plastic rear window into place. Thinking about it years later made Peter Schutz laugh. In his earliest days, he had questioned and probed every detail, understanding that Porsches were too expensive and had too many quality problems.

"So we built a convertible, raised the price twenty percent, and created a whole new set of quality problems we never had before. Now we had tops that leaked and whistled and didn't fit. And people ate it up!" Porsche sold the 1983 base 911SC for DM 57,800, $29,950 in the United States. The cabriolet went for DM 64,500 in Europe, 10 percent more, but $34,450 in America, 20 percent higher. (The Targa straddled the middle at DM 60,620D and $31,450.) The cabrio weighed about 15 kilograms (33 pounds) less than the Targa, and with its top raised and windows closed, it provided a better drag coefficient at 0.395 than the SC coupe at 0.40. Porsche test drivers reached a top speed of 235 kilometers per hour (146 miles per hour) in a closed convertible on Volkswagen's high-speed Ehra-Lessien test track, making it the world's fastest production convertible.

Other than the startling breath of fresh life that the cabriolet signaled, Porsche made few dramatic changes to its 911SC between the B Program 1981 models, C Program 1982 cars, and D Program 1983 911SC. For 1981 engineers raised the engine compression ratio to 9.8:1 from 8.6:1, introduced new pistons that increased horsepower and reduced fuel consumption, and revised valve timing. The changes required drivers to use premium fuel, consistent with the rest of Porsches products. The company extended the corrosion warranty to seven years and added the entire car body to its coverage. Externally, turn signal repeater lights appeared on front fenders (on non-U.S. cars).

▲ (Above) Porsche first showed the convertible in March 1972 at the Geneva International Motor Show. Designer and engineer Gerhard Schröder created a collapsible roof using three bows and rigid panels of aluminum.

▲ (Top right) The Cabriolet weighed 1,160 kilograms, 2,552 pounds. The company sold it for 64,500DM and $31,450 at U.S. dealers. Porsche manufactured 4,096 the first year.

For 1982 engineers revised mounting points for camshaft chain sprockets and added an internal voltage regulator to a new 1,050-watt alternator, necessitated by the car's increased electrical requirements. Tony Lapine's stylists, ever subduing bright chrome, changed the Fuchs wheel centers to black powder coat. Non-U.S. models got a standard headlight-cleaning system and better temperature control and modulation on air conditioner–equipped cars. A final option, M473, offered the front and rear spoilers from the 3.3-liter Turbo on SC coupes, Targas, and cabriolets.

For Turbo owners (though the car still was not available in the United States or Japan), a new option appeared for 1983: the M505 *flachtbau*, a flat- or slant-nose front end reminiscent of the racing 935s, for an extra DM 38,340 ($15,035 at the time). Zuffenhausen assembly did this conversion on raw body shells, allowing Porsche to provide buyers the same seven-year corrosion warranty. A few updates to Turbo engines increased torque and made

YEAR	1980-1983
DESIGNATION	911SC

SPECIFICATIONS

MODEL AVAILABILITY	Coupe, Targa, Cabriolet in 1983	SUSPENSION FRONT	Independent, wishbones, MacPherson struts, longitudinal torsion bars, hydraulic double-action shock absorbers, anti-roll bar	BORE AND STROKE	95x70.4mm/3.74x2.77 inches
WHEELBASE	2272mm/89.4 inches			HORSEPOWER	188@5500rpm (930/09 - 1980)
LENGTH	4291mm/168.9 inches				180@5500rpm (930/07 – 1980-1983 US)
WIDTH	1652mm/65.0 inches				
HEIGHT	1320mm/52.0 inches				204@5900rpm (930/10 – 1981-1983)
WEIGHT	1160kg/2552 pounds	SUSPENSION REAR	Independent, light alloy semi-trailing arms, transverse torsion bars, hydraulic double-action shock absorbers, anti-roll bar		
BASE PRICE	$25,797 coupe/$27,445 Targa (1980)			TORQUE	195lb-ft@4200rpm (930/09 - 1980)
	$25,294 Cabriolet (1983)	BRAKES	Ventilated discs, 2-piston cast iron fixed calipers		175lb-ft@4200rpm (930/07 – 1980-1983 US)
TRACK FRONT	1369mm/53.9 inches				
TRACK REAR	1379mm/54.3 inches	ENGINE TYPE	Horizontally-opposed DOHC six cylinder Typ 930/09 (1980)		197lb-ft@4200rpm (930/10 – 1981-1983)
WHEELS FRONT	6.0Jx15			COMPRESSION	8.5:1 (930/09, 930/07)
WHEELS REAR	7.0Jx15		Horizontally-opposed DOHC six cylinder Typ 930/10 (1981-1983)		9.8:1 (903/10)
TIRES FRONT	185/70VR15		Horizontally-opposed DOHC six cylinder Typ 930/07 for US 50 states and Canada, w/oxygen sensor, catalytic converter (1980-1983)	FUEL DELIVERY	Bosch K-Jetronic fuel injection
TIRES REAR	215/60VR15			FINAL DRIVE AXLE RATIO	3.875
CONSTRUCTION	Unitized welded steel			TOP SPEED	140mph
		ENGINE DISPLACEMENT	2994cc/183.2CID	PRODUCTION	21,109 coupes, 13,240 Targas, 4,096 Cabriolets all years

▲ **1984 911 SC/RS**
Known internally as Typ 954, these were the first of Porsche's Gruppe B Evolution rally cars. Porsche assembled 20 of the 1,057-kilogram (2,325-pound) cars.

them quieter in countries such with strict noise standards, such as Switzerland. An optional Performance Kit, again not available to U.S. or Japanese customers, took horsepower to 330 DIN for an additional DM 20,975 ($8,225 in 1983 dollars). This kit installed a larger turbocharger, a more efficient intercooler, and a four-pipe exhaust system.

Ekkehard Zimmerman, the man behind the radical appearance of the Kremer brothers' 935K series racers, took advantage of Porsche's marketing decision and the Kremers' racing successes to begin selling his own dp935 in 1983 out of his facilities in Cologne. He produced dozens of 935 Kremer lookalikes for the autobahns of Germany and, with modifications, the interstates of America. His technicians modified engines to develop 400 DIN horsepower, pushing his coupes and cabriolets to 280 kilometers per hour (175 miles per hour) and to 62 miles per hour in 4.8 seconds. Zimmerman's conversions added a third to half the price of a production Turbo.

More subtle changes marked Porsche's U.S.-bound normally aspirated cars for 1983. Back in 1975, DOT regulations had forced Porsche and other makers to raise ride heights to comply with 5-mile-per-hour bumper impact laws. For 1983 Porsche lowered the car back to rest-of-the-world height, improving the appearance and handling of the car. On the instrument panel, the speedometer read to 160 miles per hour. For American customers and buyers around the world, these incremental improvements made it clear that Porsche was moving ahead with the 911.

▲ (Top) Part of the Porsche development cycle for new race cars was to run 1,000 kilometers, 625 miles, over the "Belgian Road," known for deep potholes and big bumps. The company charged 188,100DM ($63,980 at the time) for these cars. *Photograph courtesy Porsche Archive*

▲ (Bottom) Strictly conceived for international rallying, these cars received a tuned 2,994cc (182.6-cubic-inch) SC Typ 930/18 engine that developed 255 horsepower at 7,000 rpm. Depending on gearing, they accelerated from 0 to 100 kilometers per hour in 5.3 second and reached 255 kilometers (155 miles) per hour.

YEAR	1980–1985
DESIGNATION	911 Turbo
SPECIFICATIONS	
MODEL AVAILABILITY	Coupe, Slant Nose Coupe
WHEELBASE	2272mm/89.4 inches
LENGTH	4291mm/168.9 inches
WIDTH	1775mm/69.9 inches
HEIGHT	1310mm/52.4 inches
WEIGHT	1300kg/2860 pounds
BASE PRICE	$71,871 complete
TRACK FRONT	1432mm/56.4 inches
TRACK REAR	1501mm/59.1 inches
WHEELS FRONT	7.0Jx16
WHEELS REAR	8.0Jx16
TIRES FRONT	205/55VR16
TIRES REAR	225/50VR16
CONSTRUCTION	Unitized welded steel
SUSPENSION FRONT	Independent, wishbones, MacPherson struts, longitudinal torsion bars, gas-filled double-action shock absorbers, anti-roll bar
SUSPENSION REAR	Independent, light alloy semi-trailing arms, transverse torsion bars, gas-filled double-action shock absorbers, anti-roll bar
BRAKES	Ventilated, drilled discs, 4-piston aluminum calipers
ENGINE TYPE	Horizontally opposed DOHC six-cylinder Typ 930/60
	Typ 930/66 from 1983–1988; Typ 930/60 S for slant nose
ENGINE DISPLACEMENT	3299cc/201.3CID
BORE AND STROKE	97x74.4mm/3.82x2.93 inches
HORSEPOWER	300@5500rpm (Typ 930/60 and 930/66)
	330@5750rpm (Typ 930/60 S)
TORQUE	304lb-ft@4000rpm (Typ 930/60) 317lb-ft@4000rpm (Typ 930/66) 344lb-ft@4000rpm (Typ 930/60 S)
COMPRESSION	7.0:1
FUEL DELIVERY	Bosch K-Jetronic fuel injection, turbocharger, intercooler
FINAL DRIVE AXLE RATIO	4.22
TOP SPEED	170mph
PRODUCTION	14,476 total Turbo production 1978–1988; 948 Slant Nose

1984–1986 **911 CARRERA**
1986–1988 **TYP 959**
1987–1989 **911 CARRERA**
1987 **RUF CTR "YELLOW BIRD"**

CHAPTER 4

THE SECOND GENERATION CONTINUES 1984–1989

1984-1986 911 CARRERA

For its 1984 model year, Porsche introduced the Carrera 3.2, or simply the 911 Carrera, the E Program second-generation 911. The number referred to its new 3.2-liter (193.0-cubic-inch) engine. The Carrera name reminded customers of Porsche's long history with the Mexican road race. That Peter Schutz and others inside Porsche decided to brand the entire new line with the name raised eyebrows and consternation inside the company and out. Carrera models had been the premium editions, cars with the highest performance capabilities. Was Porsche diluting the name? Or was the company sending another message?

Porsche sports cars always had been expensive. International currency exchanges affected prices and revenues at Porsche more than other carmakers, since half its production went to the United States. When the dollar held strength against the deutsche mark, prices did not change in Germany, but Porsches became a better bargain to American buyers. In the early 1980s, the deutsche mark gained strength over the dollar, and prices in U.S. dealerships crept up. Peter Schutz advocated the Carrera name as a way to tell customers not only that Porsche continued to offer the 911 but also that Porsche conceived and executed these cars at the Carrera-level of design, engineering, and performance.

It started with the engine, now 3,164cc total displacement, the increase accomplished by adopting the longer stroke of the Turbo, at 74.4mm. Bore remained 95mm as it had been in the SC models. Bosch introduced

◀ The Porsche crest first appeared in 1952, but not on the nose of car until 1957. The family name runs across the top. Below is the background shield of the province of Wurttemberg where Stuttgart is located. In the center is the official badge of Stuttgart.

▲ The new 911 Carrera, introduced for 1984 model year, incorporated fog lamps into the lower valance as standard equipment. Porsche sold the coupe for 61,950DM at the factory and $31,960 in U.S. showrooms.

its Digital Motor Electronics (DME) Motronics 2 engine management technology, coupled with the latest L-Jetronic injection system, to improve fuel economy, boost performance, and decrease exhaust emissions. The engine developed 231 DIN horsepower at 5,900 rpm. Exhaust emissions regulations in the United States and Japan influenced engine changes, which among other things reduced compression from 10:3 to 9.5:1 for these two markets. Running unleaded regular fuel, these engines developed 207 SAE net horsepower. The Bosch electronics for these engines utilized a heated oxygen sensor to regulate fuel-air mixture. Both 3.2-liter normally aspirated engines and 3.3-liter turbocharged ones incorporated new oil-fed camshaft drive tensioners.

The 3.3-liter Turbo remained forbidden fruit to American and Japanese buyers. Both nations enacted emissions and impact-safety legislation that Porsche and other manufacturers chose not to meet on models they were carrying over. This opened another door to entrepreneurs. A "gray market," catering to desires for the most desirable models, sprung up in 1980, soon after Porsche ceased Turbo exports to U.S. and Canadian buyers. A loophole (or shady interpretation of the regulations) encouraged mechanics and body shop operators to convert non-U.S.-specification models acquired as used cars throughout Europe. The most responsible of these operators fitted exhaust gas recirculators, air pumps, and catalytic converters to the engines. They removed European bumpers and mounted 5-mile-per-hour impact bumpers on the front and 2.5-mile-per-hour bumpers on the rear of the cars. Depending on the honesty and capability of the shop, and the desire and gullibility of the buyers, operators charged anywhere from $6,000 to $20,000, plus the cost of the car for the conversion. Sometimes the safety work was not done, and many of these cars ran poorly. (Porsche had no exclusive on these questionable conversions. Ferrari, BMW, and Mercedes-Benz buyers also suffered the sleight of hand of those with marginal skills and questionable motives.)

▶ Few differences marked the 1983 911SC coupe interior from the new 1984 Carrera model. The new car weighed 1,160 kilograms, 2,552 pounds. European cars accelerated from 0 to 100 kilometers per hour in 6.1 seconds.

▲ **1985 Carrera Turbo Look**
Optional at further cost, Carrera buyers could add the widened doorsills and rear fender ventilation available for M505 slant nose Turbos. Prices were the same as for the Turbo.

◀ The 3,164cc (193-cubic-inch) Typ 930/21 flat six developed 207 horsepower at 5,900 rpm for U.S., Canadian, and Japanese buyers and 231 horsepower for the rest of the world from the Typ 930/20 engine.

▲ The clean lines of F. A. Porsche's first 1964 901s still rang clear 20 years later although design chief Tony Lapine banished brushed aluminum and chrome, toning everything down with matte black.

▲ Inside the doors, new side-impact door beams protected driver and passenger in the event of accident. Turbo Look models remained popular in the United States because Porsche still did not sell them in America.

▲ 1986 Carrera Turbo Look
Porsche introduced new seats for 1985. These leather-covered Sport Seats provided independent electric front and rear height and angle adjustment. Seat backs adjusted manually.

▶ The Turbo Look option not only made the Carrera look more aggressive, but it also greatly improved handling due to its wider stance. The company offered the option for Cabriolets and Targas beginning in 1985.

▲ 1986 Carrera Cabriolet
While not so much a consideration in desert environments, Porsche extended its rust warranty to 10 years beginning in 1986. It covered paint for three years and the overall vehicle for two years.

▲ The company made the leather-wrapped four-spoke steering wheel standard equipment for 1985. Speedometers for American models returned to a more realistic top speed.

▶ Larger fresh air vents filled the dashboard. Porsche lowered the seat 20mm (0.8 inches) to allow tall drivers more headroom. A new interior temperature sensor allowed even the Cabriolet to utilize automatic heat control.

▲ Porsche made its sport shock absorbers available to Cabriolet buyers for the first time. The catalytic converter–equipped U.S. Carreras accelerated from 0 to 100 kilometers per hour in 6.5 seconds.

▶ The Carrera's soft top operated manually with a zip-in plastic rear window. Carrera Cabriolets sold for 82,000DM, $36,450 in the United States.

YEAR 1984–1986
DESIGNATION 911 Carrera

SPECIFICATIONS

MODEL AVAILABILITY	Coupe, Targa, Cabriolet	TIRES REAR	215/50VR16	BORE AND STROKE	95x70.4mm/3.74x2.77 inches
WHEELBASE	2272mm/89.4 inches	CONSTRUCTION	Unitized welded steel	HORSEPOWER	231@5900rpm (930/20)
LENGTH	4291mm/168.9 inches	SUSPENSION FRONT	Independent, wishbones, MacPherson struts, longitudinal torsion bars, gas-filled double-action shock absorbers, anti-roll bar		207@5900rpm (930/21)
WIDTH	1775mm/69.9 inches			TORQUE	209lb-ft@4800rpm (930/20)
HEIGHT	1310mm/51.6 inches				
WEIGHT	1210kg/2662 pounds				192lb-ft@4800rpm (930/21)
BASE PRICE	$31,175 coupe (1984)	SUSPENSION REAR	Independent, light alloy semi-trailing arms, transverse torsion bars, gas-filled double-action shock absorbers, anti-roll bar	COMPRESSION	10.3:1 (930/20) — 9.5:1 (930/21)
	$35,070 Turbo-look coupe (1984)			FUEL DELIVERY	Bosch LE-Jetronic fuel injection
	$34,609 Turbo-look coupe (1985)				
	$34,670 cabriolet (1986)	BRAKES	Ventilated, drilled discs, 4-piston aluminum calipers	FINAL DRIVE AXLE RATIO	3.875
TRACK FRONT	1369mm/53.9 inches				
TRACK REAR	1379mm/54.3 inches	ENGINE TYPE	Horizontally opposed DOHC six-cylinder Typ 930/20 (930/21 for US, Canada, Japan)	TOP SPEED	152mph
WHEELS FRONT	7.0Jx16			PRODUCTION	35,571 coupes, 14,486 Targas, 19,987 Cabriolets all years
WHEELS REAR	8.0Jx16				
TIRES FRONT	205/55VR16	ENGINE DISPLACEMENT	3164cccc/193.1CID		

▲ This was the view many competitors got of the 959s. In 1986, the rally saw 488 cars start and just 68 complete the 14,000-kilometer (8,700-mile) event 22 days later. The car weighed 1,260 kilograms (2,772 pounds.) Porsche produced just three of these racers.

1986-1988 TYP 959

During this time, Helmuth Bott and his staff were hard at work on what became the ultimate "not for U.S. buyers" Porsche. The company showed the design study, a widened and winged, startlingly pearlescent white, all-wheel drive coupe dubbed Gruppe B on its stand at Frankfurt in September 1983. It began life with the internal designation Typ 953 as a competition car for the 1984 season. Two years later, it had evolved into the limited-production masterpiece called the Typ 959.

If Bott believed all-wheel drive belonged under a Porsche 911, Schutz understood it from development, promotion, and sales perspectives.

As Norbert Singer's 1978 Moby Dick had grown from carefully reading and interpreting FIA race car regulations, so did the competition Typ 953. The FIA's new Gruppe B category, inaugurated in 1982, provided an umbrella under which manufacturers could create a race car prior to series production, rather than the reverse, in which homologation by creating production cars preceded racing legalization. This new category incorporated automobiles from the previous groups 4 through 7, which were closed two-seaters with at least 200 assembled during the previous 12 months.

Audi's all-wheel-drive Quattros, which Ferdinand Piëch's engineers had created for international rallies, regularly won individual events and season championships. They demonstrated the superiority of this technology. Porsche might create an entry, but it was venturing late into a field crowded with aggressive competitors. Worse, homologation required Porsche to manufacture many more of these all-wheel-drive cars than Bott or Schutz were ready to authorize. Another series, however, desert rallies, carried few rules and offered several advantages.

Porsche's customer racing coordinator, Jürgen Barth, and other engineers in the Competition Department urged Bott to develop a mid-engine race car based on the 914 for Gruppe B. Barth knew that 80 percent of his customers raced on paved circuits but just 20 percent ran rallies. In 1982 Porsche had introduced its mid-engine Gruppe C race car, the Typ 956, which won its debut at Le Mans and took first again in 1983. So Bott disagreed.

"We do so many mid-engine cars. We cannot learn anything," he explained. He worried about the competition: Was anyone else constructing a mid-engine Gruppe B car? If Porsche offered the sole mid-engine entry, would there even be a class?

"If you have to do a car," Bott explained, "which you have to build two hundred times, you can also build one thousand times. If we build a Gruppe B car, let's have a look at the future of the Nine Eleven." In early January, he formalized his concept, starting with a pure competition version and then a series production model utilizing technology developed for racing but contained in a customer-friendly package.

The already legendary Paris–Dakar Rally seemed a suitable venue to test a new vehicle. "Rally" was the term organizers used for a high-speed race through the desert. Many-time Porsche racer Jacky Ickx had won the 1983 Paris–Dakar in a factory-prepared Mercedes-Benz 280GE Gelandewagen, a kind of all-wheel-drive sport utility vehicle. Ickx brought valuable experience to Porsche's program. Bott completed his prototype all-wheel-drive 911 in late 1981. Engineers continually tested it and upgraded it. To prepare for the 1984 rally, they took the prototype to Africa for Ickx to test.

"After the second day," Bott recalled, "one of the front axles broke." Mechanics removed the other front axle, and Ickx ran it as a rear-wheel-drive car to the next stop. "And that was really a key thing," Bott continued. "With four-wheel

▲ There was nothing simple about the 959, even in the driver/navigator compartment. Computers and manual controls varied differential lock and traction control as well as calculated elapsed time and distance to the next checkpoint.

▲ Porsche's unsuccessful 1985 entry had used the same platform but a 3.2-liter 230-horsepower normally aspirated flat six. For 1986, the company returned with a sequential twin-turbocharged 2,849cc (173.8-cubic-inch) 400-horsepower computerized Typ 959/50 engine that put Porsches across the finish line first, second, and sixth.

▲ Porsche assembled 113 of these coupes for 1986. Its 959/50 engine with 2,850cc (173.9-cubic-inch) displacement developed 450 horsepower at 6,500 rpm. Porsche quoted a top speed of 315 kilometers (197 miles) per hour.

drive, the car handling was perfect. You could exactly drive between the dunes, allowing only fifteen centimeters [about 6 inches] clearance. And then when it became the rear-driven car, we had to allow two meters left and right. The center of gravity moved around all the time on the sand." The next morning, mechanics installed replacement front axles. Ickx again was able to position the car precisely, driving much faster.

"We thought, 'if a car is so much better under bad conditions, then you must feel it on the dry road.'

"You see, our concept with the Nine Eleven has always been that it's an all-around car. With very few changes you can drive a desert rally and then go to the racetrack at Le Mans. To show people, without changing the concept, this Nine Eleven is capable of completely different things.

"There is really a very big love from our customers for this Nine Eleven. So we thought, 'let's see if there is anything against our building this car for the next ten years, fifteen years.' It was a goal, a task much greater than to build a race car."

Schutz's indelible marker line wrapping abruptly around the corner in Bott's office had moved beyond gesture and statement. It inspired Porsche's engineers and designers with a greater challenge: The new line suggested the 911 could go on seemingly forever. Bott wondered if it could and how it might evolve.

He appointed Manfred Bantle to direct the engineering staff as they pushed every technology available—electronics, tires, suspension hardware, engine fuel delivery, ignition, exhaust, transmissions. Yet Bott understood Porsche, the company: "While its head may look into the clouds, its feet remain on the ground. Our research department was not looking twenty or thirty years ahead," he said. "In that time, things were changing so quickly. So what we do instead is to look ten years ahead. And what's possible in ten we fulfill in two and we end up eight years ahead of the others. With our small numbers of production, we are able to do this. These are the secrets."

Bott's concept was typically broad: There would be a desert rally car, the Typ 959. He and Bantle would spin off a series of 20 cars as the road-racing Typ 960. Bott's belief that if Porsche could assemble 200 then it could manufacture 1,000 took his concept to the next level, the Typ 961 next-generation Porsche Turbo. (Internal numbers shifted; the road race car became the Typ 961, and the Turbo went into development as the Typ 965.)

Bantle understood four-wheel drive. Porsche had acquired an English Jensen sports car in 1965. Four-wheel-drive pioneer Harry Ferguson had collaborated with Allan and Richard Jensen, mating Ferguson's FF all-wheel-drive system to Jensen's C-V8 coupe. When the car arrived at Weissach, Bantle analyzed its engineering and estimated its potential.

The result, Porsche's Typ 953, went straight to the winner's circle in 1984, with René Metge and Dominique Lemoyne finishing the 12,000-kilometer (7,500-mile) event in first place. Bantle biased his all-wheel drive with 60 percent at the rear, 40 at the front. Drivers could change the proportions manually and lock the differential in or out as needed. Engineers fitted the car bodies with large Carrera flat wings and aluminum, Kevlar, or fiberglass panels for front and rear deck lids and doors. Side windows were Plexiglas. They drilled large holes in the thick Kevlar belly pan to reduce weight.

The 13,000-kilometer event in 1985 yielded the opposite results. None of the three cars finished. For 1986 Jacky Ickx, managing the Paris–Dakar team effort, went to the rally start well prepared, equipped, and supported. Bott and Bantle delivered three new 959s with single turbocharger engines tuned to develop 370 DIN horsepower at 6,500 rpm. These cars reached 210 kilometers (about 131 miles) per hour along the fastest stretches. A new front axle system electronically decreased oil pressure from the drive clutches on those high-speed rear-drive sessions. Touching the brake pedal instantly restored 19 atmospheres, or 275 (psi), to the clutches for engine braking through the front axles.

Injuries to drivers and fierce storms along the route whittled the starting list of 500 entrants down to 80 finishers. Ickx's teammate Metge won as he had in 1984, and Jacky finished fourth. Rules required support personnel to accompany the entrants, so Porsche's third entry was Roland Kussmaul and Kendrick Ünger, Weissach's two development engineers. They finished sixth overall.

Porsche designer Dick Soderberg created the looks of this 911 of the future. He had designed Moby Dick with Norbert Singer, and he understood what racing rules allowed. Soderberg's Gruppe B show car advanced Moby's most recognizable styling characteristics. Racing regulations and the prohibitive costs of new body stamping dies limited the designer. He carried over the 911's existing doors and roof, as he had done with Moby. Bott and Bantle hoped the 959 would emerge from design studios and wind tunnels without spoilers or wings. But when budgets prohibited a new roofline, the rear wing was essential. With his elegant form, Soderberg created a design icon that blended seamlessly into his widened rear fenders. Its effects were more than aesthetic; the wing effectively lengthened and flattened the roofline, manipulating air flowing over the body as though it still hugged the contours. He flush-mounted the windshield; all of this helped reduce the car's drag coefficient from production figures of 0.39 to 0.32.

Bott and Bantle had hoped to introduce the production version in 1984 as a 1985 model year product. But the advanced technologies engineers developed for the car sometimes went beyond production suppliers' capabilities. DuPont had to reformulate its aramid-fiber Kevlar to form curved body panels. Bilstein invented electronically controllable "active" shocks that not only absorbed surface undulations but also lowered the car as speed increased, while still allowing drivers manual override. Dunlop developed a tire capable of long runs in excess of

320 kilometers—200 miles—per hour that could also run flat for 80 kilometers (50 miles) at much slower speeds. WABCO Westinghouse perfected anti-lock braking systems (ABS) that worked with four-wheel drive; front tires rotated faster than rears, outside tires moved more quickly through turns, any of these characteristics could change on mixed driving surfaces, and, most importantly, none of this necessarily required braking. Bosch uprated its DME Motronics computers to monitor acceleration, braking, steering, tire traction, and suspension loading 200 times every second.

As complicated as each of these new systems and materials was, it was the 959's engine that delayed the car longer than any other feature. Engine designer Hans Mezger had introduced water-cooled cylinder heads atop air-cooled cylinder walls on Moby Dick. Next-generation engines powered Typ 956 racers to countless victories. Dual overhead camshafts opened and closed four valves per cylinder. Mezger initially drove these with single-row timing chains. But odd, unpredicted torque loads and chain resonances broke them. However, to accommodate double-row chains, production engine chief Paul Hensler and his engineers had to redesign the engine block. While they were at it, they conceived a progressive water-cooled twin turbocharger configuration that responded to driver inputs the way primary and secondary jets in carburetors had done for decades. At normal engine loads, one turbo took care of engine intake needs. At higher speeds or greater acceleration, the second one came into effect. Early 959 prototypes resembled ducting and plumbing nightmares. When they finished their work, engineers coaxed 450 DIN horsepower out of the 2,849cc (173.8-cubic-inch) engine (with 95mm bore and 65mm stroke).

Porsche was not alone exploring the future of its cars at this time. As Bantle and his engineers developed new systems, they encountered delays from suppliers who were working at capacity testing or manufacturing parts for other carmakers. No other manufacturer was looking 10 years out, and rigid confidentiality agreements kept Porsche's secrets. Still, new materials and electronics gave everyone new ideas. The delays Porsche encountered with its 959 prototypes led to another problem: The European Union tightened emissions standards for model year 1985 cars. No one at Porsche worried about this when the car was a 1984 desert racer. However, as Bott and Schutz pushed the road car introduction back, new standards became serious issues.

By 1985 Group B rallying had suffered from its successes and its excesses. Cars, some weighing barely 820 kilograms (1,800 pounds) and with 600-horsepower engines, had injured spectators and drivers in high-speed crashes. The highly visible series was losing its appeal. Porsche began to question its participation with its 959, so its role as a vehicle offering the ultimate in street performance became primary. Peter Schutz and Helmuth Bott believed in the car and what it would show the public about Porsche's capability. Other manufacturers offered attractive, powerful, fast automobiles. But none approached the end-of-century technology the 959 presented.

Schutz knew Porsche could afford this development. The favorable DM-to-dollar exchange rate had helped increase 1984 company profits by 33 percent over 1983. Currency forecasts suggested this rate would raise 1985 profits 30 percent higher still. This was unsustainable economic growth, but few observers perceived it as overinflated. In May 1985, Schutz authorized

◀ The 959 "Komfort" model was intended for ultra-high-speed grand touring with exceptional stability and road feel. The six-speed gearbox included an ultra-low creeper gear for getting out of sand or snow.

▲ The 959's most significant design elements were not only the wide stance but also the integrated rear wing. Designer Dick Soderberg had to contend with aerodynamic effects without modifying the roofline, so this treatment "tricked" the wind to behave as if the roof were longer.

limited production for the 959. The following September, to display its wonders at the Frankfurt show, Bott sectioned a prototype down the middle and displayed it alongside a complete car on the stand. Deliveries began in August 1986.

Peter Schutz presided over Porsche as its U.S. sales swelled to nearly two-thirds of total output. The exchange rate, which had seen American dollars averaging 2.37 deutsche marks in 1983 and 2.85 in 1984, surged to 3.17 on the first day of trading in 1985 and averaged 2.94 for the year. By year-end, however, a new trend was clear, and it was downward bound. January 1986 trading opened at 2.44, and by the next New Year, it had plummeted to 1.92. Model year 1985 had set production records, with 54,458 cars manufactured. For 1986 the total for all Porsche models slipped slightly to 52,939; sales lagged behind economic reality.

E Program 1984 Carrera production was 14,309 cars. An M491 option turned the standard Carrera coupe into a Turbo-look wide body with a front spoiler and rear whale tail. M491 also incorporated Turbo suspension, brakes, wheels, and tires.

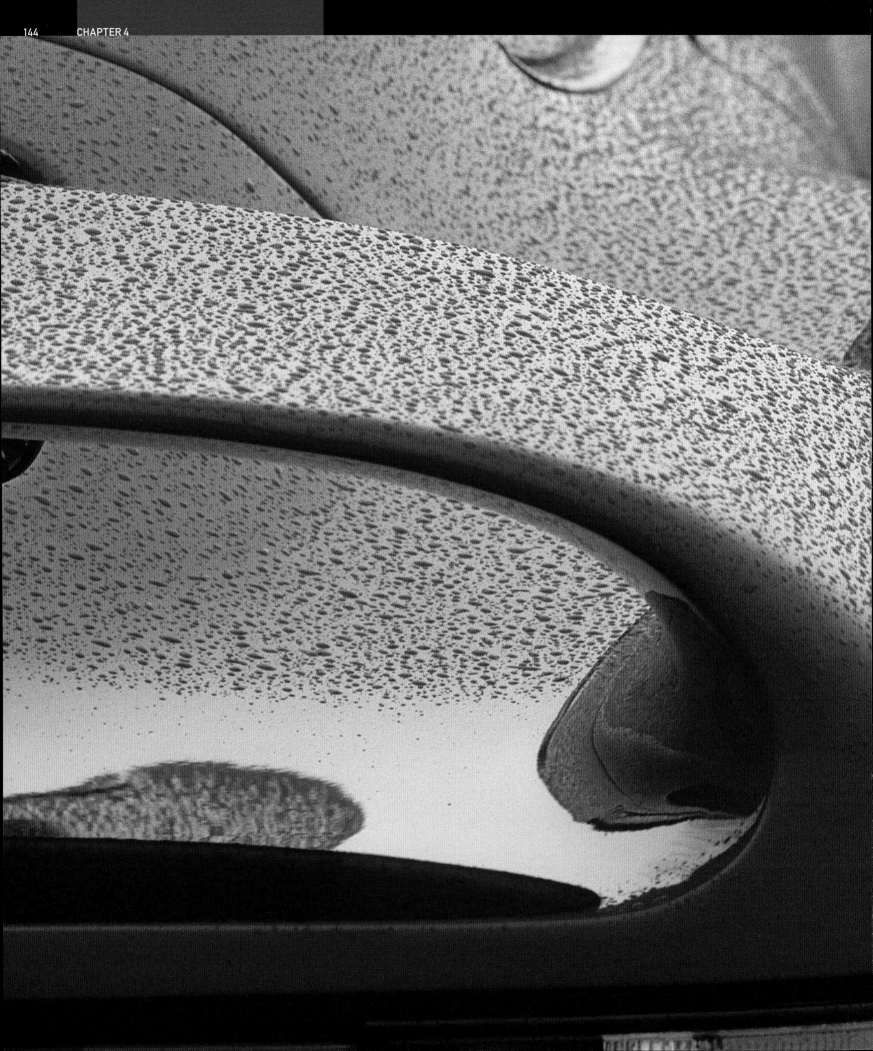

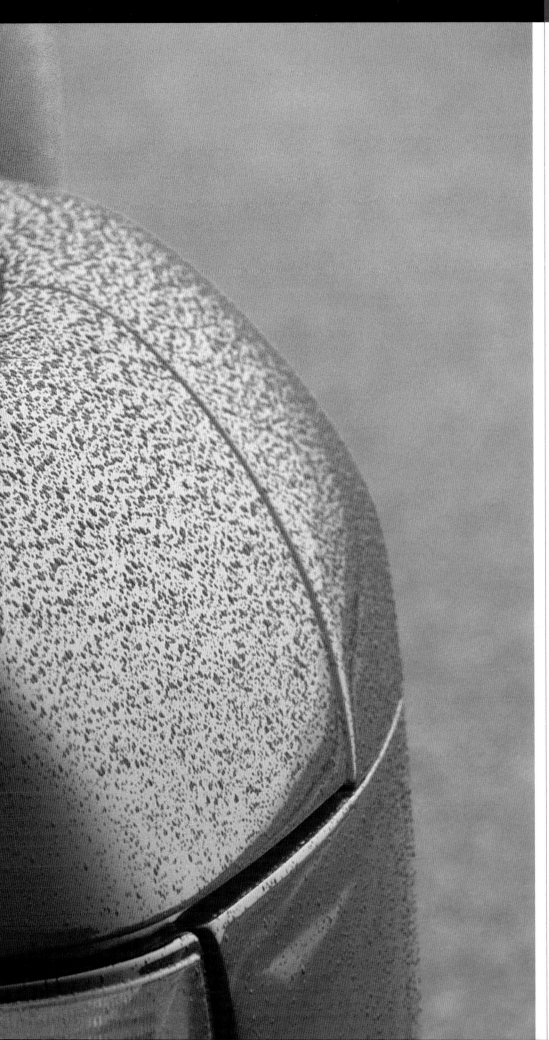

While this wider stance improved handling, the increased frontal area decreased top speed slightly. Another option mounted a front spoiler and rear wing on the standard body. This lessened front and rear lift at high speed and slightly increased the car's top speed capability.

While Weissach developed the Typ 959, it still needed a car suitable for 1984 Group B events for its customers. Jürgen Barth's engineers introduced a run of 20 "evolution" cars based on the previous generation 911SC platform to meet homologation requirements; these were known as the 911 SC/RS. Weissach fitted 3-liter engines with 935 cylinder heads using Kugelfischer mechanical fuel injection. Forged pistons and 10.3:1 compression helped it develop 255 DIN horsepower at 7,000 rpm. Rules allowed the car to run as light as 960 kilograms (2,116 pounds). The Racing Department reached this weight by fitting plastic bumpers, aluminum panels, and thin side glass to the Turbo body (with its suspension and brakes). Inside it removed the rear seats and trim, the glove box lid, and the clock, and it replaced stock seats with barely padded competition buckets. Barth's engineers timed the car from 0 to 100 kilometers (62 miles) per hour in 5 seconds and from 0 to 160 kilometers (100 miles) per hour in 11.7 seconds. Porsche sold the car for DM 188,100 (about $66,000 at the time).

◀ Porsche used DuPont Aramid and fiberglass-reinforced epoxy resin to form the rear body, the roof, and the rear wing. The 959 accelerated from 0 to 100 kilometers per hour in 3.9 seconds, and 200 kilometers (124 miles) per hour arrived in 14.3 seconds.

For F Program 1985 production, the company expanded the M491 option to encompass the Targa and cabriolet, providing Turbo-look bodies for the full Carrera line. The open cars had required an extra year while engineers developed structural reinforcement to cope with the Turbo suspension's higher cornering potential. In the interiors of all Carreras, Porsche introduced power-adjustable seats and a new four-spoke leather-covered steering wheel. To thwart vandalism, the company embedded the radio antenna in the 911's windshield. Weissach improved the shifter linkage and switched to Boge dual-tube gas-pressurized shock absorbers all around. Turbo buyers (still not in the United States or Japan) got heated seats, central electronic locking, thicker anti-roll bars front and rear, and a larger master brake cylinder. Manufacturing totals for 1985 model year Carreras and Turbos reached 13,041, including 34 M506 steel slant-nose Turbos among the 1,148 turbocharged cars delivered to rest-of-the-world customers.

Porsche upgraded warranties on G Program cars for the 1986 model year, stretching coverage to 10 years for corrosion, 3 years on paint, and 2 years/unlimited mileage for the mechanicals and body. Inside the cars, engineers and stylists redesigned the instrument panel and dashboard, revising the switch-gear and enlarging air vents. They also lowered the front seats 20mm (0.8 inches) to provide taller drivers greater headroom and incorporated sliding covers over makeup mirrors in the sun visors. A new interior temperature sensor gave 911 cabriolets automatic climate control, which Targas and coupes had incorporated since 1984. Underneath the car, thicker anti-roll bars improved ride and handling, and Porsche offered sports shock absorbers on the cabriolet as well as Targa and coupe models.

Frankfurt-based b+b, best known for engine upgrades and outrageous interiors, had introduced digital instrument panels while continuing with such in-car extras as rear-seat champagne coolers and color television sets. By 1985 demand for such options was waning, and in 1986 b+b and its 45 employees left the business.

While catalytic converters were mandatory on U.S.-destined 911s, Porsche offered the emission-control device as an option to rest-of-the-world customers as well. As with the American Carreras, the converters reduced horsepower from 231 to 207 DIN. Acceleration suffered slightly; the car took 6.5 seconds to reach 100 kilometers per hour.

The Turbo returned to U.S. dealerships for 1986. Bosch modified and reprogrammed ignition and fuel mixture controls, and Weissach added the converter, oxygen sensor, and secondary air injection system. The engine developed 282 SAE net horsepower at 5,500 rpm. The car sold in the United States for $48,000 and for DM 119,000 at the factory.

As Turbos reached American customers, Porsche offered two versions of the 959 to European buyers at the Frankfurt show. The Sport model came with cloth seats and an integrated roll bar. The Comfort version offered heated leather seats and a full leather interior. Each sold for DM 420,000. This was about $140,000 at introduction, but this figure rose as the dollar fell. By the time deliveries began later in 1986 and early 1987, the U.S. price had reached nearly $240,000. Porsche asked for a DM 50,000 deposit (initially $16,500) from each buyer. By March 1986, it had accepted 250 deposits, ensuring a full 200-unit sellout. The company relied on purchasers to drop out, lose patience, or be disqualified. An increasingly reliable cause for disqualification was U.S. residency. After struggling to meet DOT and EPA safety and emission standards for the regular Turbo, Schutz and Bott concluded that they could sell out 959 production in the rest of the world without going through the extra costs, challenges, and delays to obtain U.S. certification. They had, however, accepted 50 American deposits.

Porsche assembled 16 prototypes and 21 pilot production 959s in 1985. The tiny shops dedicated to these cars, including a former bakery space across the street from Zuffenhausen assembly, turned out 113 "production" versions in 1987 and 179 in 1988. (In 1992 the factory manufactured eight final cars from extra parts left over from 1980s production. With these, total output reached 337.)

The desire American buyers felt for a 959 led to hard work and heartbreak. Racer and Pennsylvania Porsche dealer Al Holbert worked with a new company-owned distribution organization called Porsche Cars North America (PCNA). Together they intended to import 30 cars, stripped of all interior appointments and listed on shipping and customs paperwork as racing cars, not legal for public road use. When the first eight vehicles arrived, the EPA inspected them and refused to allow the 959 into the country. PCNA had to crush the cars or return them to Germany. Through the next two years, Porsche sold about 16 to American buyers with the proviso that the cars had to remain outside the United States. For buyers with business operations throughout the world, this proved manageable. And over a number of years, this arrangement allowed at least a dozen cars to find their way into the

▲ Compared to the contemporary Carrera at 4,291mm length (168.9 inches), the 959 was slightly shorter at 4,260mm (167.7 inches). The complicated 959 weighed 1,450 kilograms (3,190 pounds).

◄ Sport versions, both U.S. and rest of the world, came equipped with cloth sport seats, five-point harnesses, and roll cages. Sport and Komfort versions sold for 420,000DM, roughly $233,000 at the time.

States, not including the single U.S. 959 Sport that the EPA and DOT did allow.

Otis Chandler, *Los Angeles Times* newspaper publisher, 935 racer, and vintage car collector, had established a museum outside L.A. to house and display his collection. He obtained a waiver allowing him to show the car at his museum and to transport it to shows and exhibitions elsewhere. He took delivery in August 1988. Over the next decade, a number of owners applied for waivers to possess and drive their cars legally. In 2001 and 2002, Congress enacted two pieces of legislation "legalizing" 959s. An industry dedicated to converting 959 engines, bumpers, seat belt systems, door locks, and lights to meet U.S. regulations arose, but as of 2008, cars at 20 years of age no longer require these changes. A flurry of buyers acquired 959s soon after that.

Porsche's Typ 961 road racer debuted in 1986 at Le Mans. It ran in the IMSA GTX (experimental) class because the FIA had no four-wheel-drive racing series. And because Porsche's 959 production had not begun yet, the car was not homologated as a Group B entrant. The 961 differed considerably from the desert rally car. Instead of positioning it high off the ground to avoid rocks and ditches, engineer Ekkehard Kiefer dropped the car 10mm (0.4 inches) below production 959 ride height. He created slots and ducts across the nose to cool the turbochargers, engine, and brakes. Plastic replaced several steel body panels, dropping the weight to 1,150 kilograms (2,535 pounds), still heavy for a race car due to its all-wheel-drive mechanisms. Development engineer Roland Kussmaul made other tweaks, including enlarging the integrated rear wing.

According to historian Glen Smale, the FIA's conditions for allowing the 961 to race at Le Mans forced Race Engine and Chassis Department chief engineer Günther Steckkönig to develop single-

▶ Total production of 959s reached 292 cars, 30 of which were intended for U.S. buyers. U.S. DOT and EPA inspectors denied them entry.

▼ Dick Soderberg, who designed the 959, also created the 1978 935-78 Moby Dick for Porsche's racing department. He adopted the extra-wide wheelwells and large rear wing, updating it for a road car. *Photograph © 2011 Dave Wendt*

THE SECOND GENERATION CONTINUES 1984–1989

piece water-cooled heads for each bank of three cylinders. Hans Mezger's racing engineers tuned 640 DIN horsepower out of the 2.8-liter (170.8-cubic-inch) twin-turbocharged engine by retuning the Bosch Motronic system, enhancing the fuel mixture intercooling, running open exhaust, and increasing turbo boost from the 0.8 bar (12 psi) on production models to 3.25 bar (47 psi) for Le Mans. While Porsche originally intended to run 19-inch wheels and tires, according to Jürgen Barth, Dunlop offered a wider selection of compounds in 17-inch sizes, and "this size was well represented in the U.S. where they intended to race the car" in other IMSA Gruppe B races. For Le Mans, Porsche put three-time Paris–Dakar winner (and five-time Le Mans competitor) René Metge in the car with Claude Baillot-Lena. Twenty-four hours later, after few flaws in an otherwise uneventful race, the two drivers finished seventh overall in the all-white race car. (They followed six other Porsches to the checkered flag.) It was the first time an all-wheel-drive race car competed at Le Mans.

Four months later, Kees Nierop joined Steckkönig as co-driver for a three-hour race at Daytona as a way of introducing the car to North America, where Porsche hoped IMSA would develop a multiseason four-wheel-drive racing series. But the heavy car placed huge loads on the tire sidewalls while up on the high banking, and the 961 finished only 24th overall.

Its second appearance at Le Mans, in June 1987, did not end well either. Engineers coaxed another 10 horsepower out of the engine and switched to 19-inch wheels and tires front and rear (after hoped-for recognition from IMSA and U.S. appearances failed to materialize). But driver Kees Nierop's accidental downshift spun the car just before 9:00 a.m. on Sunday morning, and he crashed head on into the Armco. The violence of the impact spilled oil onto the engine, and the back of the car burned badly. Nierop was running 10th at the time. He escaped unhurt. Porsche retired the 961 from competition.

YEAR	1986–1988
DESIGNATION	959
SPECIFICATIONS	
MODEL AVAILABILITY	Four-wheel drive Coupe, Komfort (Luxury) or Sport
WHEELBASE	2290mm/90.2 inches
LENGTH	4260mm/167.7 inches
WIDTH	1840mm/72.4 inches
HEIGHT	1490mm/59.6 inches
WEIGHT	1450kg/3190 pounds Komfort; 1350 kg/2,970 pounds Sport
BASE PRICE	$217,500
TRACK FRONT	1514mm/59.6 inches
TRACK REAR	1525mm/60.0 inches
WHEELS FRONT	8.0Jx17
WHEELS REAR	9.0Jx17
TIRES FRONT	235/45ZR17
TIRES REAR	255/40ZR17
CONSTRUCTION	Composite, galzanized steel, aluminum, plastic
SUSPENSION FRONT	Independent, double wishbones, twin coil-over variable rate/height shock absorbers per wheel, anti-roll bar
SUSPENSION REAR	Independent, double wishbones, single coil-over variable rate/height shock absorber per wheel, anti-roll bar
BRAKES	Ventilated, drilled discs, 4-piston aluminum calipers
ENGINE TYPE	Horizontally opposed DOHC six-cylinder Typ 959/50, water-cooled cylinder heads
ENGINE DISPLACEMENT	2850cc/173.9CID
BORE AND STROKE	95x67mm /3.74x2.64 inches
HORSEPOWER	450@6500rpm
TORQUE	369lb-ft@5000rpm
COMPRESSION	8.3:1
FUEL DELIVERY	Bosch DME, Sequential twin turbochargers, intercoolers
FINAL DRIVE AXLE RATIO	4.125:1
TOP SPEED	196mph
PRODUCTION	16 prototypes, 21 pre-production cars, 163 Komfort, 30 US Sport

1987-1989 911 CARRERA

H Program cars for 1987 saw significant improvements. The most noticeable for Carrera buyers was a new hydraulically operated clutch and five-speed transmission, the G50 model from Getrag. It introduced a new shift pattern, relocating reverse from right-and-down to left-and-up. Getrag's linkages and Borg-Warner's synchromesh system made shifting smoother and easier. Engine tweaks increased output of all catalyst-equipped Carrera engines from 207 to 217 DIN horsepower, still trailing behind the uncatalyzed engines producing 231 horsepower.

Optional on Carreras and standard on Turbos, two new power supplies appeared on the cabriolets. Through a system of motors and flexible and mechanical drives, owners could raise or lower the cloth top in 20 seconds from the comfort of their seats. This complex system incorporated 22 pivot points and 13 control levers, roof bows, and frames. It operated only when the vehicle was stationary. The plastic polycarbonate rear window still had to be zippered into place.

After several years as an option, the M505/506 steel slant-nose treatment for the Turbo became a regular production model designation, the Turbo SE in the United Kingdom and the 930S in the United States, starting in March 1987. Porsche manufactured a total of 591 of these bodies in coupe, Targa, and cabriolet configuration, with distribution around the world. Coupes sold for DN 202,330, roughly $93,240 at the factory at the time.

Beginning with 1987 and continuing through 1989, Porsche offered a Club Sport version of the Carrera coupe. The cars, with 231 DIN horsepower, were not available in the United States and sold originally for DM 80,500, roughly $44,700

▲ (Top) Porsche improved door seals and added discrete rain gutters to the Targa roof to better protect occupants in inclement weather. The Targa sold for 80,880DM in Germany and $33,450 in the United States.

▲ (Bottom) To the great happiness of American customers, U.S.-destined 911s now looked like those for the rest of the world, better fared-into the fender lines. Front tires grew from 185/70-15 to 195/65-15.

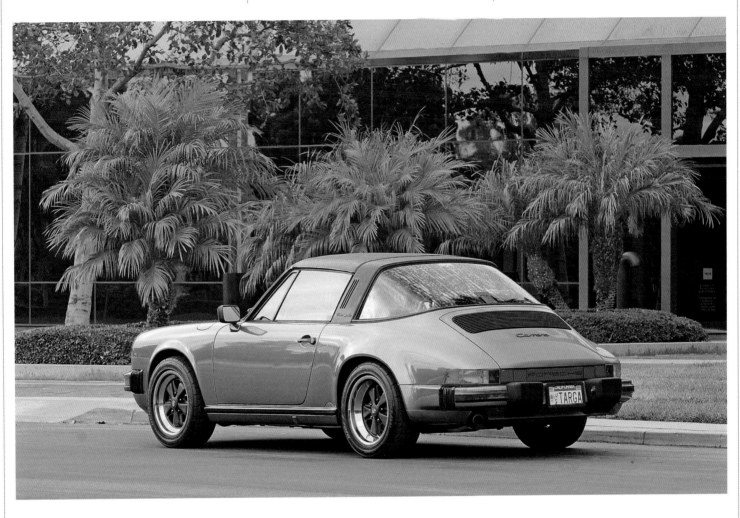

▲ The most significant change to 1987 models lay ahead of the engine in the new G50 transmission with Borg-Warner synchromesh and hydraulic clutch. Engine output of catalyst-equipped Typ 930/25 engines increased to 217 horsepower at 5,900 rpm.

at the factory. In addition to 189 coupes over two years, Zuffenhausen produced a single Targa Club Sport. All 190 came with a stiffer suspension, the fixed whale-tail rear wing, a front spoiler, and a Club Sport graphic on the nose of the car.

Just after Porsche introduced its J Program 1988 model year to the public, the world's financial picture changed considerably. Currency exchange rates already had affected Porsche's profits, especially from sales in the United States. The dollar brought 1.8 deutsche marks on average until soon after October 19, 1987. That day on Wall Street, the Dow Jones Industrial Average plummeted 508 points, losing nearly 23 percent of its value. Through 1987 Porsche production slipped to 48,520 cars. U.S. sales that year dipped to 23,632 units, below half the total for the first time in more than a decade. Numbers fell further in 1988, when U.S. sales of all models reached just 15,737. Peter Schutz slashed production.

Porsche, hoping to precipitate buyer interest, celebrated the milestone of 250,000 911s manufactured on June 3, 1987. For 1988, on the 25th anniversary of the 911, Porsche released a commemorative model in Diamond Blue metallic paint and loaded with extra features as standard equipment, including a passenger power seat, central locking capabilities, and a headlight- and more extensive windshield-washing system. The cars were available as coupe, Targa, or cabriolet models. Seats, in silver blue crushed leather, had headrests embossed with Ferry Porsche's signature. The company assembled 875 anniversary cars; 250 remained in Germany, 300 came to the United States, and the rest sold into other markets.

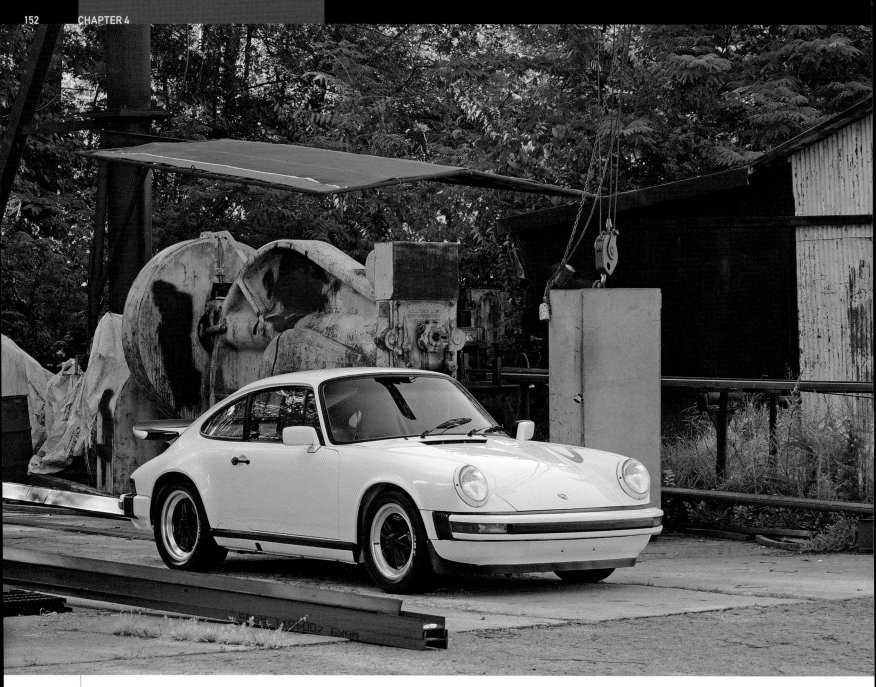

▲ Throughout model years 1987, 1988, and 1989, Porsche manufactured 189 of these Club Sport Carreras. The company deleted anything not necessary for competition, including standard Carrera fog lights.

▶ The only "badge" on the Club Sport was this graphic along the front fender. Porsche sold the 1988 model for 82,000DM, roughly $46,590, at the factory. The company manufactured a few Club Sport Targa models.

▲ Porsche fitted the simplest whale tail rear wing to the car. Horsepower output remained unchanged at 231 (making it a non-U.S. product), but the Typ 930/20 engine revved more quickly due to lighter intake valves.

◀ Rear seats came out, as did the radio, the covers over the door pocket, sun visors, and power window systems. Compared with the 1,210-kilogram (2,662-pound) Carrera coupe, the Club sport weighed 1,160 kilograms (2,552 pounds).

▲ In the final year of the 911 Carrera 3.2, Porsche introduced a Speedster version, 35 years after bringing out the original 356 model. The company produced them both in narrow body and Turbo Look versions.

▶ There was no difference between the instrument panel and seats of the Speedster and the Carrera Cabriolet models. Porsche fitted a slightly lower windshield in an aluminum frame that owners could remove from the car.

▲ The most significant styling statement of the car came from its distinctive rear "humps," part of the plastic rear hatch that tilted up for manual roof access. The car was strictly a two-seater.

◀ Porsche manufactured 2,103 Turbo Look Speedsters. They sold for 110,000DM in Germany and $65,480 in the United States.

YEAR	1987-1989
DESIGNATION	911 Carrera

SPECIFICATIONS

MODEL AVAILABILITY	Coupe, Clubsport Coupe, Targa, Cabriolet;	TIRES FRONT	195/65VR15 (205/55ZR16 Speedsters)
	Speedster and Speedster Turbo Look (1989 only)	TIRES REAR	215/60VR15 (245/45ZR16 Speedsters)
WHEELBASE	2272mm/89.4 inches	CONSTRUCTION	Unitized welded steel
LENGTH	4291mm/168.9 inches	SUSPENSION FRONT	Independent, wishbones, MacPherson struts, Longitudinal torsion bars, gas-filled double-action shock absorbers, anti-roll bar
WIDTH	1652mm/65.0 inches		
	1775mm/69.9 inches (Speedster Turbo Look)		
HEIGHT	1320mm/52.0 inches	SUSPENSION REAR	Independent, light alloy semi-trailing arms, transverse torsion bars, gas-filled double-action shock absorbers, anti-roll bar
	1220mm/48.0 inches (both Speedsters)		
WEIGHT	1260kg/2772 pounds (coupe, Targa, Cabrio)		
		BRAKES	Ventilated, discs, 2-piston cast iron calipers
	1160kg/2552 pounds (Clubsport)	ENGINE TYPE	Horizontally opposed DOHC six-cylinder Typ 930/20
	1210kg/2662 pounds (Speedster)		
			(930/21 for US, Canada, Japan)
	1290kg/2838 pounds (Speedster Turbo Look)	ENGINE DISPLACEMENT	3164cccc/193.1CID
		BORE AND STROKE	95x70.4mm/3.74x2.77 inches
BASE PRICE	$33,450 (1987 Targa)	HORSEPOWER	231@5900rpm (930/20) 207@5900rpm (930/21)
	$33,745 (1988 Clubsport)		
	$65,480 (1989 Speedster)	TORQUE	209lb-ft@4800rpm (930/20) 192lb-ft@4800rpm (930/21)
TRACK FRONT	1398mm/55.0 inches		
	1372mm/54.0 inches (Speedster)	COMPRESSION	10.3:1 (930/20) – 9.5:1 (930/21)
		FUEL DELIVERY	Bosch LE-Jetronic fuel injection
	1432mm/56.4 inches (Speedster Turbo Look)		
		FINAL DRIVE AXLE RATIO	3.44
TRACK REAR	1405mm/55.3 inches (Speedster)		
		TOP SPEED	152mph
	1492mm/58.7 inches (Speedster Turbo Look)	PRODUCTION	35,571 coupes, 14,486 Targas, 19,987 Cabriolets all years
WHEELS FRONT	7.0Jx15 – (7.0Jx16 Speedster Turbo Look)		189 Clubsport coupes, 1 Clubsport Targa; 171 Speedsters;
WHEELS REAR	8.0Jx15 – (9.0Jx16 Speedster Turbo Look)		2,103 Speedster Turbo Look

▲ The 930 S used the Performance Kit option, taking the 3,299cc (201.2-cubic-inch) turbocharged/intercooled Typ 930/60S engine to 330 horsepower at 5,750 rpm. Several modifications through Porsche's *Sonderwunsch*, "Special Wishes," department brought output up considerably higher.

▲ **1990 930 S Turbo Cabriolet**
Company engineers and designers got involved with this project once everyone recognized that this was the final 3.3-liter Turbo to be produced.

▲ Starting with model year 1987, Porsche referred to the Slant Nose Turbo models as the 930 S, offering them in coupe, Targa, and Cabriolet models. Louvers atop the front wheels were an extra cost option, 2,185DM, $1,162.

◀ A sensor registered rainfall, lowered side windows, raised and locked the cloth top, closed the windows, re-armed the alarm, and sent the owner a pager message.

▲ Alois Ruf fitted the "Group C Turbo Ruf" coupe with his twin-turbocharged 3.4-liter flat six, developing 469 horsepower at 5,950 rpm. When *Road & Track* magazine set out to find the fastest car on earth, this CTR reached 342 kilometers (213 miles) per hour on the Nardo circuit in Italy. *Photograph courtesy Ruf Automobile GmbH*

1987 RUF CTR "YELLOW BIRD"

To establish another milestone, Alois Ruf accepted a challenge from *Road & Track* magazine to determine the fastest car on earth. He created what looked like a standard Carrera, painted it yellow, and designated it CTR 001. Despite its brilliant color, its looks were severely understated, part of Ruf's shrewd philosophy. The car weighed just 1,173 kilograms (2,580 pounds), with 469 DIN horsepower on tap. Ruf and his staff modified the body to be more slippery, the suspension and brakes to handle exceptional speeds, and, of course, the engine to produce them. The magazine nicknamed it Yellow Bird, and when its trial against a dozen challengers was complete, Ruf's narrow-body turbo had reached 211 miles (338 kilometers) per hour, making it the fastest of the lot.

He assembled another 28 CTRs and converted 25 (or so) more 3.2 Carrera coupes into the 469-horsepower turbocharged coupes.

Porsche again used the 1987 Frankfurt show to test customer response and excite buyers by showing a "design study" called the 911 Speedster. Weighing about 70 kilograms (154 pounds) less than the production cabriolet, the Speedster was strictly a two-seater, with an unpadded, hand-operated cloth top hidden beneath a molded plastic tonneau cover. According to Jürgen Barth, this was to be the first of three variations on the Speedster that also included a removable hardtop version and a Club Sport version with a tiny windshield that was strictly meant for road course competition.

Porsche's Speedster, fitted with the standard Carrera 3.2-liter engine of 207 SAE horsepower in the United States and 231 DIN elsewhere, became part of the 1989 model year K Program. Coupe, Targa, and cabriolet models continued largely unchanged, except for a new alarm/immobilizer system that placed light-emitting diodes (LEDs) in the door lock buttons as standard equipment. Engineers worked out the kinks of mating the Turbo's power and torque to the Getrag transmission, and the G50 five-speed became standard for 1989. Porsche manufactured 2,103 Turbo-look Speedsters and 171 standard-body models for worldwide markets. But its previously largest market, the United States, was hemorrhaging. Sales of all models in America reached only 9,479 cars.

Over the course of its life, from 1984 through 1989, Porsche had raised the price of the 3.2 Carrera coupe from DM 61,950 (roughly $21,737 at the time) on introduction to DM 86,000 ($45,745 in 1989 dollars). U.S. buyers saw prices increase from $31,960 to $51,204. Comparing dollars to deutsche marks is not exactly pricing apples against oranges. The difference in those figures reflected the healthy profit Porsche enjoyed from favorable exchanges. The more than doubling in dollar values of home-market prices represented Schutz's efforts to maintain costs, trim overhead (he interrupted production for a week or longer each month starting in 1988 to keep employees on the payroll and the company alive), and to return some profit to owners.

Unfortunately, neither Schutz nor Helmuth Bott, who had conceived it, was at Porsche to see the 1989 Carrera Speedster come to the market. Bott had supervised the 959, and while it garnered Porsche thousands of pages of newspaper and magazine attention, crediting the company with advancing the state of the art in automotive technology, the entire project cost the company nearly half a billion deutsche marks. In 1988 dollars at 1.76 to the deutsche mark, that figure of roughly $240 million amounted to roughly $720,000 per car, about three times the delivery price. Peter Schutz shouldered the responsibility for high costs and overruns. Blame did not come from the owners, but it resonated with several of them and among outside shareholders, who lost 30 percent of their share value as effects of the 1987 market crash rippled around the world.

Peter Schutz endured two tough years. During the last one, for most of 1987, his wife, Sheila, lived in Florida, where she built a business restoring, renovating, and reselling homes. She visited him regularly, but she found Germany's male-dominated society uninviting to an independent outsider with ambition. Schutz suspected that those who blamed him for economic downturns welcomed the chance to end his contract a year early. In early December, Porsche's Supervisory Board accepted his resignation. By mid-January 1988, he was at home in Florida.

Helmuth Bott, having lost his champion in the abrupt and dramatic change of management, held on at Weissach until late 1988 to see off the last of the 959s. When they were gone, except for a few spare parts to build a few final cars a few years later, he retired. He was 63. At the beginning of his career, he had tamed the handling of the 356s. At the end, he established a new automotive category, the super car. His 959 became the ultimate driving machine, the target for other manufacturers who dared to dream.

YEAR	**1987**
DESIGNATION	**Ruf CTR "Yellow Bird"**
SPECIFICATIONS	
MODEL AVAILABILITY	Coupe
WHEELBASE	2272mm/89.4 inches
LENGTH	4251mm/167.4 inches
WIDTH	1652mm/65.0 inches
HEIGHT	1270mm/50.0 inches
WEIGHT	1150kg/2530 pounds
BASE PRICE	$160,000
TRACK FRONT	1384mm/54.5 inches
TRACK REAR	1417mm/55.6 inches
WHEELS FRONT	7.0Jx17
WHEELS REAR	10.0Jx17
TIRES FRONT	215/45VR17
TIRES REAR	255/40VR17
CONSTRUCTION	Unitized welded steel
SUSPENSION FRONT	Independent, wishbones, MacPherson struts. Longitudinal torsion bars, gas-filled double-action shock absorbers, anti-roll bar
SUSPENSION REAR	Independent, light alloy semi-trailing arms, transverse torsion bars, gas-filled double-action shock absorbers, anti-roll bar
BRAKES	Ventilated, drilled discs, 4-piston aluminum calipers
ENGINE TYPE	Horizontally opposed DOHC six-cylinder
ENGINE DISPLACEMENT	3366cc/205.4CID
BORE AND STROKE	98x74.4mm/3.86x2.93 inches
HORSEPOWER	469@5950rpm (930/20)
TORQUE	408lb-ft@5100rpm (930/20)
COMPRESSION	7.5:1
FUEL DELIVERY	Bosch Motronic, Twin turbochargers, twin intercoolers
FINAL DRIVE AXLE RATIO	3.44:1
TOP SPEED	213mph
PRODUCTION	1 Yellow Bird

▲ The CTR engine, with 98mm (3.86-inch) bore and 74.4mm (2.93-inch) stroke, displaced 3,366cc (205.3 cubic inches). The car, nicknamed Yellow Bird for its bright color during a dull grey day photo shoot, weighed 1,170 kilograms (2,580 pounds). Cars sold for 288,000DM, roughly $160,000. *Photograph courtesy Ruf Automobile GmbH*

1989–1993 **911 CARRERA**
1991–1992 **911 TURBO**
1990–1991 **CARRERA 4 LIGHTWEIGHT**
1992–1994 **CARRERA TURBO LOOK (AMERICA ROADSTER), CARRERA RS, CARRERA RS AMERICA, TURBO SLANT NOSE, AND SPEEDSTER**
1993 **CARRERA RS AND RSR 3.8**

CHAPTER 5

THE THIRD GENERATION APPEARS 1989–1994

1989-1993 911 CARRERA

Helmuth Bott's schedule in 1984 had been full of the future, planning half a decade and further ahead. While Porsche just had introduced the 3.2 Carrera, he already was anticipating the 911's next generation. Tests with all-wheel drive for desert racing suggested to him how the technology might aid average drivers on public roads in any conditions. When he chatted with Peter Schutz, he saw three possible avenues the 911 could take as it headed off the production chart on the wall. One scenario introduced the Typ 964 as an all-wheel-drive model supplementing a carryover 3.2 Carrera. The second gave the Carrera a facelift, mounting it on new rear-wheel- and all-wheel-drive platforms. Finally, he considered designing and engineering a unitized body—which functioned as chassis—and mounting this on rear- and all-wheel drive running gear.

In 1984 Porsche had profited from the currency exchange rate of 3.5 deutsche marks to the dollar. Bott used some of the proceeds to construct a wind tunnel at Weissach. The company needed new paint facilities. He assigned a young production engineer, Wendelin Wiedeking, who had arrived in 1983, to supervise its construction. Bott next asked him to design and manage construction of the new body assembly plant, Werke V. This building linked the paint shop and final assembly by a bridge high above Schwieberdinger Strasse. A conveyor transported welded raw bodies from one building to the other high above traffic. The advantageous exchange rate allowed Porsche to

◀ **1994 Turbo S 3.6 Slant Nose**
The company manufactured just 76 of these coupes at a price of 290,000DM, $174,698. The car weighed 1,470 kilograms, 3,234 pounds. Acceleration from 0 to 100 kilometers per hour took 4.7 seconds. Top speed was 280 kilometers (174 miles) per hour.

▲ **1989 Carrera 4 Coupe**
Porsche's new 911 produced a coefficient of drag (Cd) of 0.32, a great improvement over the Carrera 3.2 model with its figure of 0.59. A completely flat undertray helped a great deal, as did carefully sculpted rain gutters along the windshield and over the doors.

undertake these huge projects without borrowing any money.

From this position of financial independence, and inspired by drawings and models of the 959, the Supervisory Board selected Bott's option for the unitized body with new mechanicals. But then it took a step back, permitting stylists Wolfgang Möbius and Dick Soderberg only to update the 911 body. The already classic stovepipe fenders and headlights, as well as the large roof panel, were untouchable. Porsche had the resources to do something as radical as the 959, but the board worried that customers would not embrace radical change.

Bott's concept specification book (*lastenheft*) for the Typ 964 addressed a new concern: the coefficient of drag (Cd). Porsche's 911 looked sleek, but its Cd measured 0.395, a figure that embarrassed company engineers and designers when compared to new Ford and Audi sedans, at 0.29 and 0.30 respectively. Soderberg's wide 959 came in at 0.31, and Bott targeted 0.32 for the new 911. Soderberg and Möbius flush-mounted front bumpers, carefully adjusted front valences, and installed a belly pan under the nose. This protected the front differential and driveshaft from impact and eliminated turbulence as well.

▲ The new car, known as the Typ 964 internally, was 85 percent new from the Carrera 3.2. This Typ M64/01 engine developed 250 horsepower at 6,100 rpm. It sold for 116,600DM in Germany, $69,500 in the United States.

◀ A completely new 3,600cc (219.6-cubic-inch) flat six with 100mm (3.94-inch) bore and 76.4mm (3.00-inch) bore powered the new 911. Each cylinder had two spark plugs firing the fuel/air mixture.

▲ 1990 Carrera 2 Coupe
The 964 wheelbase measured 2,272mm (89.5 inches), same as the Carrera 3.2. It measured 4,250mm long (167.3 inches), while the previous Carrera was 41mm longer at 4,291 (168.9 inches).

The 911 body form resembled an airfoil, and the slope from the peak of the roof to the taillights caused the rear to lift at high speeds. Tilman Brodbeck's *bürzel* (ducktail) temporarily reduced the problem, but it was time for a new approach. Engineers and stylists developed an electrically operated spoiler that rose from the rear deck lid at 80 kilometers (50 miles) per hour. Its location and length reduced rear lift to zero.

For the complicated drivetrain of the all-wheel-drive 964, Bott and production car engineer Fritz Bezner maintained the 911's traditional rear weight bias, settling on 59 percent on the rear wheels. They allocated 69 percent of the power to the rear, leaving 31 percent to front tires. Bott said his goal was "to provide customers with handling characteristics that felt familiar to them, and were similar to rear-drive Nine Elevens, but with the benefit of additional traction in poor conditions."

Oftentimes in developing a new car, engineers face good news/bad news choices. Weissach refers to these as target conflicts, and the 964 presented many of them. The need for a driveshaft forward of the engine put Porsche's trusty torsion bar rear springs in the way. The shaft had to go right through the transverse tube that anchored them. That tube contributed considerable stiffness to the platform. The new rear suspension, with coil springs as at the front, coupled with aluminum semi-trailing arms, was a compromise.

Engine chief Paul Hensler needed to use two spark plugs for ignition to meet California's nitrogen oxide (NOx) emission standards. Twin plugs burned the fuel more efficiently from idle all the way to redline, leaving little or no unburned matter for catalysts to incinerate. Complete combustion also increased horsepower. Here was another target conflict: Two spark plugs in air-cooled heads compromised the space for cooling fins. Hoping to effect more efficient heat transfer from the cylinder heads down to the heavily finned cylinder walls, Hensler's engineers eliminated the traditional head gasket between them. This decision eventually caused leakages that subsequent engineers had to address. These kinds of target conflicts delayed the 964 engine, and the model launch slipped to 1989.

It was a heavier car, almost 20 percent more so than the 3.2 Carrera. New computers, electronics, and other features and their wiring added more than 227 kilograms (500 pounds) to the 964. The car weighed in at 1,455 kilograms (3,200 pounds). To meet Bott's performance targets, Fritz Bezner and Paul Hensler developed a new 3,506cc (213.9-cubic-inch) engine, achieved by increasing bore to 100mm. This modest increase did not move the all-wheel-drive car as quickly as the current 3.2, let alone meet Bott's goal. They lengthened the stroke 2mm as well, to 76.4mm, which yielded 3,600cc total displacement, or 219.6 cubic inches. This engine developed 250 DIN horsepower at 6,800 rpm (239 SAE net at 6,100) and 229 lb-ft of torque at 4,800 rpm. The board approved production as a 1989 model year product, and marketing named it the Carrera 4, quickly abbreviated to C4. The rear-drive version, Carrera 2 or C2, followed in early 1990.

The work it required of Weissach engineers and the delays those efforts caused pushed the car over its development budget. The year-late introduction, combined with a still-declining deutsche-mark-to-dollar exchange rate and the departure of company chairman Peter Schutz, hurt the company. The board elected a conservative and cost-conscious member named Heinz Branitzki to be chairman. Branitzki had joined Porsche's board in 1971, when the 911 was seven years old. Over the next 18 years, he watched the car evolve through subtle changes. He expected this would be the plan in 1989. In the 964 introduction press kit, he stated, "We have here the 911 for the next 25 years, the concept that will help our favorite model reach its 50th anniversary."

While everyone around him hoped the 911 would reach 50, few hoped it would be the 964 getting them there. The car was much that enthusiasts, journalists, and Weissach staff had expected. But it was less than the engineers and stylists had desired. More significantly, it cost much more than most people could imagine. Porsche boasted that 85 percent of the parts were new. While this

▲ Porsche introduced the all-wheel-drive Carrera 4 in spring 1989. Later that summer, the company brought out the rear-drive Carrera 2 model. Without the heavy front differential, the C2 weighed 1,350 kilograms (2,970 pounds), 100 kilograms less than the C4 at 1,450 kilograms, 3,190 pounds.

▲ A new interior provided owners a full-length central console from which a shorter gear lever appeared. The leather-covered seats were adjustable for height as well as rake and fore-aft travel.

was good for advertising and marketing, it raised eyebrows among insiders. The number of parts that Porsche shared among 3.2 Carrera models, the new 944 Turbo, the 928, and the 964 was so small that when Wendelin Wiedeking examined parts lists, development costs, and production revenues, he concluded that no one monitored Porsche's engineering expenses. He went so far as to accuse Helmuth Bott of destroying the company through fiscal practices that Peter Schutz had either ignored or approved of. When Wiedeking sensed his discoveries meant nothing to the board, he left Porsche. But Bott, saddened and disheartened, left within a few months as well, just before the 964 Carrera 4 launch, two years before his scheduled retirement.

When the new car reached dealers, the deutsche-mark-to-dollar exchange rate of 1.76 to 1 put the K Program 1989 964 Carrera 4 at DM 114,500 in Germany and $69,500 in the United States. With the introduction of the new car, Porsche ended production of 3.2 Carrera bodies, except for preordered Turbo models. The L Program 1990 964 C2 arrived without the C4's front half shafts, differ-

▼ One of the 964's cleverest innovations was its rear spoiler that rose from the engine deck lid automatically at 80 kilometers (50 miles) per hour. This served to reduce rear lift to zero, and it exerted downforce on the front axle.

ential, and drive shaft. This saved it 100 kilograms, or 220 pounds, and increased front luggage capacity. The Carrera 2 debuted in coupe form at DM 103,500 and $58,500 in America. Cabriolets (with standard-equipment power-operated tops) and Targas appeared on both rear-drive and all-wheel-drive platforms.

Besides significant traction differences, there were subtle distinctions between the C2 and C4 models. C2 versions used a vacuum-booster braking system with two-piston fixed caliper brakes at the rear. The C4 utilized a hydraulic brake booster. All models came with ABS as standard equipment. Porsche coupled the standard five-speed manual transmission to a dual-mass flywheel, reducing transmission rattle at low engine speeds. By the mid-1990 model year, prices had risen slightly in Germany, to DM 120,550 for the C4 coupe and DM 107,100 for the C2.

In September 1989, as Porsche introduced the rear-drive C2 models, Ferry Porsche turned 80. As had been the case on significant anniversaries in the past, the company presented him with a special one-off automobile. Few had been as radical or as startling as this gift, dubbed the Panamericana. The idea had begun when styling chief Harm Lagaay arrived at Porsche in early 1989. He found the staff he inherited idly dreaming up projects to keep themselves busy. He assigned one of the dreamers, Steve Murkett, the task of creating Ferry Porsche's birthday gift in a vehicle that had all-wheel drive and remarkable capabilities. Murkett came up with the X-Country, a dune-buggy-inspired 911 that evolved into the green roadster birthday gift.

▲ The new normally aspirated 964 was virtually as fast as the previous generation Turbo. Acceleration from 0 to 100 kilometers took 5.7 seconds. Top speed was 260 kilometers (162 miles) per hour, the same as the 1989 Turbo coupe.

▲ Porsche sold Carrera 2 Cabriolets for 127,525DM with the standard manual gearbox and 133,780DM in Germany for the Tiptronic-equipped open car. U.S. dealers charged $70,690. Acceleration from 0 to 100 kilometers took 6.6 seconds.

◄ The four-speed Tiptronic transmission allowed drivers to relax with the gearbox in full-automatic mode or slide the lever to the right and perform gear changes themselves. It added 6,000DM, about $3,600, to the price of the car.

► (Opposite) At the time Porsche introduced the rear-drive Carrera 2 models for 1990, it brought Cabriolet and Targa bodies to the full lineup. Midway through that year, the company also introduced the new Tiptronic Typ A 50/01 computer-controlled four-speed gearbox that functioned both as a full automatic or semi-manual (without a clutch pedal).

YEAR: **1989-1993**
DESIGNATION: **911 Carrera**

SPECIFICATIONS

MODEL AVAILABILITY	Coupe, Targa, Cabriolet	WHEELS REAR	8.0Jx16	ENGINE TYPE	Horizontally opposed DOHC six-cylinder Typ M64/01
WHEELBASE	2272mm/89.4 inches	TIRES FRONT	205/55ZR16	ENGINE DISPLACEMENT	3600cc/219.7CID
LENGTH	4250mm/167.3 inches	TIRES REAR	225/50ZR16 Speedsters	BORE AND STROKE	100x76.4mm/3.94x3.00 inches
WIDTH	1652mm/65.0 inches	CONSTRUCTION	Unitized welded steel	HORSEPOWER	250@6100rpm
HEIGHT	1310mm/51.6 inches	SUSPENSION FRONT	Independent, lower wishbones, MacPherson struts w/coil springs, gas-filled double-action shock absorbers, anti roll bar	TORQUE	229lb-ft@4800rpm
WEIGHT	1450kg/3190 pounds (Carrera 4)			COMPRESSION	11.3:1
	1350kg/2970 pounds (Carrera 2)			FUEL DELIVERY	Bosch DME with sequential injection
BASE PRICE	$69,500 (1989 C4 coupe)			FINAL DRIVE AXLE RATIO	3.44:1
	$58,500 (1990 C2 coupe)	SUSPENSION REAR	Independent, MacPherson struts w/coil springs, gas-filled double-action shock absorbers, anti roll bar	TOP SPEED	161mph
	$70,690 (1991 C2 Cabriolet)			PRODUCTION	13,353 coupes, 1,329 Targas, 4,802 Cabriolets from 1989 through 1993
TRACK FRONT	1380mm/54.3 inches				
TRACK REAR	1374mm/54.1 inches	BRAKES	Ventilated, discs, 4-piston aluminum calipers		
WHEELS FRONT	6.0Jx16				

▲ To accommodate the wider front and rear flared fenders of the new Turbo body, Porsche created a new doorsill molding. The car used "Carrera Cup" outside mirrors.

1991-1992 911 TURBO

One victim of Peter Schutz's and Helmuth Bott's departures was the 965 all-wheel-drive Turbo project. Designer Tony Hatter made the shapes of the 959 evolve from sketches to models to reality by adopting the rear cooling slots, nose and headlights, and integrated rear wing into his own concept for Porsche's next-generation Turbo.

Bott and Fritz Bezner endured cooling and horsepower output challenges with the 965 engines, similar to those that troubled Paul Hensler with his 964 engines. This prompted Bott and Manfred Bantle to consider alternatives. They looked into developing either a V-6 version of Porsche's newest V-8 Indy engine, with four valves per cylinder and dual overhead camshafts, or detuning a version of it. These options were viable but were time-consuming and costly. By late summer 1988, with Porsche internally in turmoil, water cooling a 3.5-liter variation of the flat six was the favored solution.

By this time, Peter Schutz had gone. Styling chief Tony Lapine, recuperating from a heart attack, was asked to stay home, and Helmuth Bott had left. Porsche hired Ulrich Bez from BMW to take over Bott's role as technical director. He immediately reexamined engine possibilities for the 965. As 911 development chief Fritz Bezner watched the debate, he began planning a fallback strategy. The Turbo was a profit leader for Porsche. Doing without one would hurt the company. He recognized that the 964 Turbo could grow from the "Sportkit" 330-horsepower version of the 3.3-liter 930 engine. It needed catalytic converters, requiring intake, exhaust, and cylinder head modifications. A month before the full management board meeting on December 20, Bez ordered Bezner to hurry development of this engine. Bez and passenger car chief engineer Horst Marchart set a target of 300 horsepower. At the meeting, the board agreed this

(CONTINUED ON PAGE 174)

▶ The new Turbo accelerated from 0 to 100 kilometers per hour in 5.0 seconds and was capable of a top speed of 270 kilometers (167 miles) per hour. It cost 183,600DM, about $110,600, at the factory.

▼ The company fitted 17-inch "Cup-Design" wheels to the Turbo. The car ran 205/50ZR17 tires on the front and 225/40ZR17 on the rear. It weighed 1,470 kilograms, 3,234 pounds.

▲ For the 1991 Porsche Carrera Cup Deutschland season, Paul Strähle's AutoSport operation in Schöndorf won the championship with their driver Roland Asch in this car. It was their third championship in four years.

◀ (Opposite) The 3,299cc (201.2-cubic-inch) Typ M30/69 engine developed 320 horsepower at 5,750 rpm. Porsche used cross-drilled rotors and four-piston fixed calipers with a reconfigured ABS to stop the car.

▲ Roland Kussmaul in the racing department created these Cup cars. He lowered suspension 55mm (2.17 inches) and fitted specially manufactured Bilstein shock absorbers. The Typ M64/01 Cup engine developed 265 horsepower at 6,100 rpm.

(CONTINUED FROM PAGE 170) was the new production 964 Turbo; and it killed the 965, sending its 16 running prototypes to crushers, cutting torches, and storage.

Stylist Tony Hatter went back to work. "We had to redo it completely," he said. "Take off all the aggressive looking hoops and things and make it a reasonable car." Instead he widened front and rear bodywork and, with the help of engineers and aerodynamicists, created a new rear wing and new oval outside mirrors, adopted from Carrera Cup racers.

Porsche started production in May 1990 for the 1991 model year in Europe. Bezner retuned the 3,299cc (201.2-cubic-inch) engine to an emissions-controlled 320 DIN horsepower at 5,750 rpm, with 332 lb-ft of torque at 4,500 rpm. His engineers adjusted the ABS and fitted cross-drilled rotors and four-piston fixed calipers on all wheels. A new limited slip differential, with variable "lockup" factors under acceleration, coasting, and braking, effectively managed the earlier Turbo's characteristic oversteer. Porsche introduced the car at DM 183,600 ($110,600 in 1991 dollars). Remarkably, the company sold the car for $95,000 in the United States.

Two more target conflicts came to resolution in a performance record for the Turbo. The U.S. Congress enacted corporate average fuel economy (CAFE) ratings. In effect, this system permitted large auto manufacturers to develop and assemble less-efficient gas guzzlers, so long as they also sold fuel-efficient models that reduced the company's fuel economy average to reach federal targets. To smaller companies such as Porsche, this situation represented a sizable hurdle. A fundamental technique to reduce fuel consumption was to slow engine speed by using higher gear ratios to achieve the same road velocity. Slower engines also ran quieter, a benefit to Swiss buyers, who encountered roadside microphones that enforced their country's strict noise regulations. However, since gearing changes did not affect peak engine speed, higher gears produced higher velocities. As a result, German and English magazines testing the new Turbo reached speeds of 274 kilometers (171 miles) per hour. While government regulatory challenges such as these taxed engineers, designers, and production chiefs at Porsche, a more modest project in the racing department that invited U.S. scrutiny took form.

Late in 1988, Jürgen Barth proposed to competition director Peter Falk that Porsche might assemble two or three 964 RSR racers based on the C2 platform using 3.4-liter engines to compete in the 24-hour endurance race at Nürburgring. Porsche sales, including customer race car sales, had fallen with the depressed exchange rates. Years earlier, Norbert Singer had given a presentation graphically proving that when Porsche raced, sales rose. Falk approved the Nürburgring cars and a second project as well, one strictly for customers, based on the 964 C4 platform and some exotic spare parts Weissach had in storage.

Historically, when Porsche developed a new-technology race car, it assembled the few examples that other competitors and the public would see. But it also ordered many sets of spare parts. In the case of the 953-959 Paris–Dakar entries, Porsche produced three cars but had 30 sets of all-wheel-drive running gear. Barth envisioned producing a car similar to the 961 but for customer use for circuit racing.

He knew all about Ferdinand Piëch's 911Rs. He had tested prototype 911 RS Carrera 2.7s and the racing RSR 2.8 and 3.0 models. FIA regulations at the time offered no series for all-wheel-drive cars to race, but Barth had seen manufacturers exert influence before. He felt certain that if Porsche pressed the point, a series and venues for races would open up. This was 1988, not 1973 or 1967, however; the element of timing made this car intriguing and its story and the lessons it produced relevant.

Throughout the world, those who survived the 1987 stock market crash and avoided bankruptcy sought new means to work their money. Successful business leaders always collected trophies, and although Porsche's dealers could not sell a $70,000 Carrera to these folks, "investors" with no interest or experience with racing were buying Porsche 908 race cars for $400,000 and selling them a month later for $700,000.

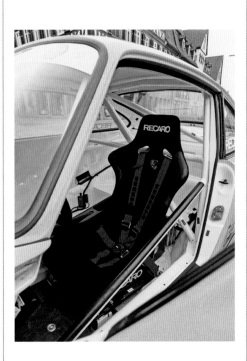

▲ Weissach gutted the interior of all creature comforts but added a rigid roll cage and Recaro racing seat. The instrument panel, however, and the steering wheel, gear shifter, and foot pedal group were stock.

▲ **1992 Turbo S Coupe**
Known internally as "the lightweight," the S came in at 1,290 kilograms, 2,838 pounds, with 381 horsepower at 5,750 rpm, due primarily to different camshafts in the Typ M30/69SL engine. The company sold the cars only in Europe for 295,000DM, roughly $189,100. *Photograph © 2011 Dieter Landenberger*

Legitimate race cars, even if they never raced, became a new investment target.

One factor made Barth's idea timely. In Washington, EPA and DOT regulators clamped down on "gray market" auto importers in 1987, enforcing rules they previously had monitored less vigorously. Red flags went up in U.S. Customs sheds, because a growing number of 1973 RS 2.7s came into the United States as race cars. These didn't look like racers; they had full interiors, and very few had roll bars. A dozen years later, this tripped up PCNA's efforts to bring in 959s and set off alarms anytime a Porsche appeared. As racing journalist Kerry Morse explained it, "You can't bring a car into the U.S. with a seventeen-digit serial number, a roll bar, and air conditioning, and call it a race car. The folks at the EPA and DOT are not stupid." Their diligence, however, contributed to the racing car investment frenzy.

▲ Porsche Cars North America proposed a Carrera Cup series for the United States, but was unable to come to agreement with any sanctioning body for 1992, so only few cars reached American racers. Most were converted back to production models.

▲ For the 1992 season engineers found another 10 horsepower in the planned M64/03 CUP USA engine to increase output to 275 at 6,100 rpm. Instead of opening a season in the United States, Porsche succeed in launching the series in Japan.

YEAR	1991-1992
DESIGNATION	911 Turbo
SPECIFICATIONS	
MODEL AVAILABILITY	Coupe
WHEELBASE	2272mm/89.4 inches
LENGTH	4250mm/167.3 inches
WIDTH	1775mm/69.9 inches
HEIGHT	1310mm/51.6 inches
WEIGHT	1470kg/3234 pounds
BASE PRICE	$95,000
TRACK FRONT	1434mm/56.5 inches
TRACK REAR	1493mm/58.8 inches
WHEELS FRONT	7.0Jx17
WHEELS REAR	9.0Jx17
TIRES FRONT	205/50ZR17
TIRES REAR	255/40ZR17
CONSTRUCTION	Unitized welded steel
SUSPENSION FRONT	Independent, lower wishbones, MacPherson struts w/coil springs, gas-filled double-action shock absorbers, anti roll bar
SUSPENSION REAR	Independent, MacPherson struts w/coil springs, gas-filled double-action shock absorbers, anti roll bar
BRAKES	Ventilated, drilled discs, 4-piston aluminum calipers
ENGINE TYPE	Horizontally opposed DOHC six-cylinder Typ M30/69
ENGINE DISPLACEMENT	3299cc/201.3CID
BORE AND STROKE	97x74.4mm/3.82x2.93 inches
HORSEPOWER	320@5750rpm
TORQUE	332lb-ft@4500rpm
COMPRESSION	7.0:1
FUEL DELIVERY	Bosch DME with sequential injection
FINAL DRIVE AXLE RATIO	3.44:1
TOP SPEED	167mph
PRODUCTION	3,660 coupes in 1991, 1992

1990-1991 CARRERA 4 LIGHTWEIGHT

Two significant race car features motivated speculators: These cars were purpose built in ultralimited quantities, and those numbers were easily verified. So these cars were importable. Morse also knew Jürgen Barth, calling on him during his frequent visits to Germany and often staying in Barth's home. On one trip in mid-1989, he stopped in and asked his friend what was new.

Barth explained that he planned to gut the 964 interior and body of all excess weight and install beefed-up running gear to create a 964 RS. He said he was considering assembling just a few—8, 10, 12—cars at around DM 200,000, roughly $110,000 at the time. Morse committed to buying one on the spot.

About a month later, Helmut Flegl, who was number two in engineering behind Ulrich Bez, questioned the project. Flegl was codeveloping the 964 Carrera Cup cars with Roland Kussmaul, and he wondered where Barth's RS fit in their programs. Barth responded to each inquiry and pressed ahead, ultimately removing close to 350 kilograms (770 pounds) of weight from the C4s. He replaced doors and front and rear deck lids with thin gauge aluminum and side windows with Plexiglas. A roll bar ran the perimeter of the stark interior, which was most remarkable for its two large knurled knobs on the instrument panel. These were not Turbo boost controls, familiar to many 935 racers, because this engine was normally aspirated. These two controls directed differential bias from front to rear and left to right.

Modified electronics and the free-flow exhaust boosted engine output to 265 DIN horsepower, up from the stock 3.6-liter engine's 250. A stainless-steel exhaust system made this one of the loudest 911 racers in history. It weighed about 1,105 kilograms (2,430 pounds) and from the outside appeared unremarkable, with neither flares nor modified bodywork. As the first example neared completion, the new Carrera 4 Leichtbau, or C4 Lightweight, earned one last important distinction.

After the dozens of failed attempts to import 1973 RS 2.7s and later 959s with 17-digit serial numbers, almost anything Porsche manufactured earned critical scrutiny from EPA, DOT, and U.S. Customs inspectors. Porsche race cars usually had six-digit serial numbers without letters. Barth got the Lightweights numbered beginning with 964001, like factory works racers rather than like customer cars with identification beginning with WPOZZZ.

Word spread among Porsche race car devotees. This was a car that might never race, but it was rare and available. The run grew to 21 cars and sold out. Number 964001 went to a collector in the United States. It and the next three emerged between September and

◀ When the goal was ultra light weight and the man in charge of the project was racer and customer racing manager Jürgen Barth, the result was a no-frills coupe weighing 1,100 kilograms, 2,200 pounds, compared to the stock C4 coupe at 1,450 kilograms, 3,190 pounds.

▶ One of its most interesting features was fully adjustable drive balance not only front to rear but left to right. The Typ M64/C4L engine developed 265 horsepower at 6,750 rpm, making the C4 lightweight an extremely potent product. Porsche assembled something like 21 of these cars.

November 1990 as 1991 models. Internal politics drove up Barth's hoped-for price of DM 225,000DM to DM 285,000 at delivery, roughly $172,000 at the time; some of Porsche's decision makers had seized the opportunity to wring extra profit from this odd market. This price was difficult to accept when race-ready 964 Carrera Cup cars, for which a series in Europe and the United States already existed, cost DM 123,000, or about $78,365 in the United States. Some people wondered aloud if 953 running gear was worth an extra $100,000. Deliveries continued through 1991, but the "collector car as investment" bubble was set to burst.

On the eve of the 1993 recession, prices for cars such as the Porsche 917K crept above $4 million, and some historic race cars sold for as much as $25 million. Prices swelled to many times life size as sharp speculators skipped to the next trend. This left hundreds of amateur players holding cars attached to ruinous loans. Barth had to send letters to buyers reminding them of their commitments.

Car 964021 left Weissach in late 1992. The C4 Lightweights made money for Porsche, as Barth's projects nearly always had done. It cost Porsche nothing to delete sound insulation and add parts that had been gathering dust. The legacy of these cars lived on in the message that special limited editions appealed to loyal enthusiasts. As it had learned with the 1973 RS, the 959, even the 1989 Carrera Speedster, Porsche understood its best customers would dig deeply to own something unique.

Far outnumbering the 70 Cup cars and the 21 Lightweights, Porsche assembled 20,666 C2 and C4 models in 1990. The M Program 1991 cars provided driver and passenger air bags. But Japanese competition ate into Porsche's market share, and the company manufactured just 13,816 cars.

YEAR	1990-1991
DESIGNATION	911 Carrera 4 Lightweight
SPECIFICATIONS	
MODEL AVAILABILITY	Coupe
WHEELBASE	2272mm/89.4 inches
LENGTH	4275mm/168.3 inches
WIDTH	1652mm/65.0 inches
HEIGHT	1255mm/49.4 inches
WEIGHT	1100kg/2420 pounds
BASE PRICE	Not available
TRACK FRONT	1420mm/55.9 inches
TRACK REAR	1530mm/60.2 inches
WHEELS FRONT	8.0Jx17 "Cup"
WHEELS REAR	9.5Jx17 "Cup"
TIRES FRONT	245/620 – 17
TIRES REAR	265/630 – 17
CONSTRUCTION	Unitized welded steel
SUSPENSION FRONT	Independent, lower wishbones, MacPherson struts w/coil springs, gas-filled double-action shock absorbers, anti roll bar
SUSPENSION REAR	Independent, MacPherson struts w/coil springs, gas-filled double-action shock absorbers, anti roll bar
BRAKES	Ventilated, drilled discs, 4-piston aluminum calipers
ENGINE TYPE	Horizontally opposed DOHC six-cylinder Typ M64/01
ENGINE DISPLACEMENT	3600cc/219.7CID
BORE AND STROKE	100x76.4mm/3.94x3.00 inches
HORSEPOWER	265@6720rpm
TORQUE	224lb-ft@6720rpm
COMPRESSION	11.3:1
FUEL DELIVERY	Bosch DME with sequential injection
FINAL DRIVE AXLE RATIO	3.44
TOP SPEED	Not available
PRODUCTION	22

1992-1994 CARRERA TURBO LOOK (AMERICA ROADSTER), CARRERA RS, CARRERA RS AMERICA, TURBO SLANT NOSE, AND SPEEDSTER

Lessons seldom get lost at Porsche. Turbo-look bodies had sold well enough on 3.2 Carrera models that marketing created similar wide bodies on 964 platforms. To commemorate a particularly significant U.S. export in 1952, the company issued a 1992 America Roadster, called the Cabriolet Turbo Look in Europe. To encourage U.S. sales, the car sold for $94,960, about $11,000 less in the States than in European markets.

Along with the America Roadster, Porsche introduced three variations on a 964 RS, with slightly more comfort than Barth's radical C4 Lightweight. Each was available only on the rear-drive platform, and each was a clear nod to the 1973 model. But here was a case where an already homologated race car, the Porsche Carrera Cup C2, inspired the customer version. The company planned initial production of 1,000 units of a car it intended to be a true dual-purpose road car/race car. To that end, Weissach engineers lowered the car 40mm (1.6 inches) and retuned springs and shocks to much firmer handling, useful for racing yet still acceptable to enthusiasts on the road. Seventeen-inch magnesium Carrera Cup wheels, non-power-assisted steering, and a brake system combining the best of Turbo and Carrera Cup pieces provided further evidence of the car's true purpose.

Inside the 964RS, Porsche used a thin carpet and headliner that barely

◀ Porsche offered the America Roadster and the European Turbo Look Cabriolet with either manual or Tiptronic transmissions. It manufactured just 250 copies. *Photograph © 2011 David Newhardt*

▲ The car was known as the Turbo Look Cabriolet in Europe; though for the United States, Porsche commemorated an earlier American-market success, the 1952 America Roadster. The company fitted Turbo brakes and bodywork but with the standard Carrera 2 movable spoiler. *Photograph © 2011 David Newhardt*

reduced engine and road noise. It eliminated rear seats and fitted racing seats and harnesses matching the car's exterior color. Gone were electric seat adjusters, outside mirrors, and other motorized amenities. Buyers could option a roll cage, six-point harness, onboard fire system, and external battery kill switch in the N-GT and Carrera Cup–eligible versions. Running 98-octane fuel, no catalytic converters, and modified pistons and cylinders, the RS engine developed 260 DIN horsepower at 6,100 rpm. Weight-saving efforts dropped the road car to 1,240 kilograms (2,730 pounds),

giving it acceleration from 0 to 100 kilometers (62 miles) per hour in 5.4 seconds and a top speed of 260 kilometers (162 miles) per hour. The lightweight N-GT (homologated as a FIA version of the Carrera Cup cars, with production finishing at 290 copies) came in at 1,220 kilograms (2,684 pounds). It sold for DM 160,000 ($102,560 at the time).

As Porsche had experienced with its 1973 RS, demand outraced supply, and the company assembled 1,053 in 1991 and another 1,352 in 1992. The company offered neither of these DM 145,450 ($91,000) models in the United States, as

they lacked both emissions controls and air bags. However, American customers enjoyed another instance of reverse price discrimination with the 1993 and 1994 RS America coupe, similar to the European touring version of the 964 RS. PCNA sold this rear-drive model for $53,900, reflecting a substantial savings over the home market version. Fitted with Carrera Cup wheels, the M030 sport suspension, and a fixed whale-tail rear spoiler, the M504 RSA provided two cloth-covered sport seats and, in a nod to American tastes, just four options, including air conditioning, a sunroof, a 90

▲ To meet market demand from the United States for the 964 RS that would not comply with U.S. DOT and EPA standards, Porsche developed the RS America based on the 1993 Carrera 2 coupe. Its most notable styling cue was its large whale tail. *Photograph © 2011 Keith Verlaque*

percent locking rear differential, and an AM/FM/cassette stereo radio. Porsche manufactured 617 in 1993 as P Program models and another 84 for 1994 under the R Program. The 1993 models had two storage bins in place of rear seats, but the 1994 cars got their rear seats and belts back. The very attractive price was nearly $10,000 less than a base Carrera 2 coupe, turning the RSA into an instant classic in the States.

Porsche returned the two-seat Speedster to its lineup for 1993 on the C2 narrow body. Similar to its 1989 predecessor, it utilized a hard plastic tonneau to cover the rear storage area and provided only a manually operated cloth top. Customers throughout the world got the Speedster with the 250-horsepower (DIN) flat six and either the standard five-speed manual or four-speed Tiptronic transmission. Electric window lifts were standard, and air conditioning was optional. The company planned production of 3,000 units, but the economy tripped up its hopes and assembly ended at 904. Rolf Sprenger's Sonderwunsch Department created about 20 Turbo Look Speedsters, reminiscent of those available in 1989.

Neither Cabriolet Turbo Looks, Speedsters, 964RS or RS America models, nor even more potent Turbos were enough to pry open the pocketbooks of cautious buyers. Porsche production for N Program model year 1992 slipped to 9,747 C2 and C4 models, plus 3,298 Turbos in all body styles. Zuffenhausen assembly turned out 1,058 Cabriolet Turbo Looks or America Roadsters in 1992 but just 321 in P Program in 1993. For the 1993 model year, the company launched its new Turbo Look bodywork as an option on

(CONTINUED ON PAGE 187)

▲ According to www.rsamerica.net, Porsche assembled 701 of these lighter-weight Carrera 2 coupes from mid-1992 into early 1994. The company offered them in five colors but encouraged customers to special order other colors for an additional $2,498.

◀ Black was the standard interior treatment, with inner door panels from the European RS missing armrests and door pockets. The company offered only four options: a sunroof, air conditioning, a limited slip differential, and a stereo cassette radio.

▲ Porsche Cars North America sold the cars for $53,900, a $10,000 price reduction from standard fully equipped C2 models. The RSA weighed 1,340 kilograms, 2,948 pounds, and accelerated from 0 to 60 miles per hour in 5.3 seconds.

▲ The Roadster and Turbo Look Cabriolet were delivered with an enhanced sound system, leather upholstery and heated seats, an on-board computer, and Automatic climate control. The normal 250 horsepower engine pushed the wider body to a top speed of 255 kilometers (155 miles) per hour. *Photograph © 2011 David Newhardt*

▲ Porsche introduced its new 3,600cc (219.6-cubic-inch) Turbo in model year 1993, as it was beginning to phase out the 964 car line. The new engine developed 360 horsepower at 5,500 rpm. A performance kit boosted horsepower to 385.

▶ The S used Porsche's M 64/50 S engine with 100mm bore and 76.4mm stroke to yield 3,600cc. The car rode on 225/40ZR18 front tires and 265/35ZR18 rears. The only option was a sunroof.

(CONTINUED FROM PAGE 183)
C4 models. This configuration appeared for 1993 as the Thirtieth Anniversary 911, fitted with the standard electrically operated rear wing. Porsche delivered the cars in Viola Metallic, Silver Metallic, or Amethyst Metallic; the full leather interior was Rubicon Gray.

Several years of working with the new 3.6-liter flat six as a normally aspirated power plant led to introduction for 1993 of the new 3.6 Turbo. Porsche's efforts with its previous 3.3-liter version had gone as far as possible, culminating with the ultralight, ultralimited-production 381-horsepower 1992 Turbo S, based on the models that had won the IMSA Supercar Championship in the United States in 1991 and 1992 (and would take it again in 1993). Porsche developed the car using tricks from its Racing Department and its 964RS program. By removing air bags, air conditioning, power seats and window lifts, and the rear window wiper, and by giving the car thinner side glass and carbon-fiber front and rear deck lids, it pared 180 kilograms (about 397 pounds) from the standard Turbo to reach 1,290 kilograms (2,844 pounds). With less weight and more power, the S accelerated from 0 to 100 kilometers per hour in 4.7 seconds and reached a top speed of 290 kilometers (180 miles) per hour. Porsche sold it for DM 295,000; this would have been about $175,000 had the car been offered in America. The company assembled 86 of these cars.

Because the S lacked emissions controls and certain U.S.-required safety features such as air bags, Porsche developed an S2 for American buyers. This regularly equipped Turbo produced a still-respectable 322 SAE horsepower. The U.S. S2 reached 60 miles per hour in 4.8 seconds and topped out at 178 miles per hour. It sold for $119,950. Production continued through introduction of the 1994 R Program.

Porsche Special Wishes also created a run of 76 slant-nose Turbos with a "Performance Kit" during 1994 model year. Retractable headlights came from 968 and 928 models, and the tuned engine developed 385 DIN horsepower at 5,750 rpm. Porsche sold these cars for DM 290,000 (roughly $179,000) to European customers.

▲ 1993/4 Carrera 2 Speedster
Unlike the 1989 version, owners could not remove the windscreen of this Speedster. The company decided not to provide passenger airbags; it was worried about the force of the inflation.

▶ Porsche introduced the Speedster as a regular production model in the narrow-body form. The *Exclusiv* department converted 15 of the cars to Turbo Look.

THE THIRD GENERATION APPEARS 1989–1994

▲ (Top) The company resurrected the hard plastic shell to cover the rear storage area and the top. Standard color selections got 17-inch Cup wheels painted in body color. For metallics, or special orders (such as this Fly Yellow), or black cars, the factory painted wheels silver.

▲ (Bottom) The company intended to manufacture 3,000, but demand ebbed at 936. It was known as Option M503. Porsche sold them for 134,000DM, $80,700. Wide-body conversion cost another 16,410DM, roughly $9,885.

YEAR	**1992–1994**
DESIGNATION	**911 Carrera Turbo Look or 911 America Roadster**
SPECIFICATIONS	
MODEL AVAILABILITY	Cabriolet
WHEELBASE	2272mm/89.4 inches
LENGTH	4250mm/167.3 inches
WIDTH	1775mm/69.9 inches
HEIGHT	1310mm/51.6 inches
WEIGHT	1420kg/3124
BASE PRICE	$104,740
TRACK FRONT	1434mm/56.4 inches
TRACK REAR	1493mm/58.8 inches
WHEELS FRONT	7.0Jx17
WHEELS REAR	9.0Jx19
TIRES FRONT	205/50ZR17
TIRES REAR	255/40ZR17
CONSTRUCTION	Unitized welded steel
SUSPENSION FRONT	Independent, lower wishbones, MacPherson struts w/coil springs, gas-filled double-action shock absorbers, anti roll bar
SUSPENSION REAR	Independent, MacPherson struts w/coil springs, gas-filled double-action shock absorbers, anti roll bar
BRAKES	Ventilated, drilled discs, 4-piston aluminum calipers
ENGINE TYPE	Horizontally opposed DOHC six-cylinder Typ M64/01
ENGINE DISPLACEMENT	3600cc/219.7CID
BORE AND STROKE	100x76.4mm/3.94x3.00 inches
HORSEPOWER	250@6100rpm
TORQUE	229lb-ft@4800rpm
COMPRESSION	11.3:1
FUEL DELIVERY	Bosch DME with sequential injection
FINAL DRIVE AXLE RATIO	3.44:1
TOP SPEED	161mph
PRODUCTION	702

1993 CARRERA RS AND RSR 3.8

More "special" still was a limited series of two-seat 911s known as the Carrera RS 3.8 and the racing version, the RSR 3.8. To meet weight restrictions, Weissach's Customer Racing Department removed power-assisted steering from the RS and RSR versions but kept and upgraded the hydraulic power brakes with ventilated and cross-drilled rotors, fronts from the Turbo S and rears from the 964 RS. An additional front spoiler increased down force, as did an adjustable biplane rear wing. Door panels and front and rear deck lids were aluminum, and the rear and side windows were thin-gauge glass. Weissach gutted the interior, leaving little more than two Recaro competition seats. The "street" version, with 300 DIN horsepower at 6,500 rpm, weighed 1,210 kilograms (2,668 pounds) and was a delight for European customers. To achieve its 3.8-liter (231.8-cubic-inch) displacement, engineers enlarged bore from 100mm to 102. The RS 2.8 was a special-order product that set its buyers back DM 225,000, about $140,000 in U.S. dollars in the spring of 1993.

The racing RSR 3.8s got off to a tremendous start in 1993. They took GT-class victories at the 24 Hours of Le Mans, at the 24-Hours of Spa in Belgium, and a 24-hour event at the Nürburgring. The normally aspirated flat six produced 350 DIN horsepower at 6,900 rpm. The racer's interior was trimmed down to a welded-in roll cage, a single Recaro seat and six-point harness, and an onboard fire system. Options included an air jacking system, center-lock wheel lugs, and an additional brake cooling system. Base price for the car was DM 270,000, nearly $168,000. In addition to endurance races in Europe, these cars competed in the Supercup series in the United States.

Back in the mid-1990 model year, Porsche introduced a new four-speed Tiptronic transmission, one of the first gearboxes offering a fully automatic gear shift and clutchless manual shifting. Racing engineers installed an innovative Porsche Doppelkupplungsgetriebe (PDK) transmission in racing 956s for development and then for competition. Helmuth Bott directed this work, hoping to use this double-clutch system in the 959 and then in subsequent production 911s. By 1986, however, a hoped-for collaboration with Audi had failed, leaving Porsche to support development costs on its own. The complex gearbox was heavy and became too expensive for use in series production cars.

Weissach engineers began work with ZF technicians on another gearbox, and as time passed, priorities came clear. They wanted a transmission that could shift automatically, but it had to respond to the way Porsche drivers operated their cars, which could be very different from how Audi or BMW or even Chevrolet drivers controlled theirs. What's more,

▲ Porsche manufactured 50 of these 1,200-kilogram (2,640-pound) GT racers. Depending on gearing, they easily exceeded 265 kilometers (165 miles) per hour. The company sold the cars for 270,000DM, roughly $166,700.

◀ In its debut year 1993, Porsche's development engineer Jürgen Barth along with Dominique Dupuy and Joel Gouhier co-drove to victory in the GT class at Le Mans. Others racked up GT wins throughout the season.

engineers wanted a manual shift function that overrode the gearbox's automatic characteristics yet would automatically shift to avoid damaging the engine or gearbox. Weissach and ZF invested four years perfecting the transmission, much of this time spent computer programming its brain. When it arrived, magazine reviewers anticipated the chance to criticize it. Instead, they found that Tiptronics provided acceleration and top-speed performance very close to the manual gearbox, and fuel economy in the same ranges. The new "Tip" found a home in one of every three C2 and C4 automobiles. Buyers willingly paid the DM 6,000 in Germany or $2,950 in the United States for the intelligent transmission.

Historians look back on this era at Porsche as a time of tumult. Heinz Branitzki, the cautious finance expert elevated to the front office to replace Peter Schutz, heralded the arrival of the 964 as the 911 for the next quarter century. Doubtlessly he hoped his engineers and designers would not need to repeat the lavish spending that had taken place under Schutz and Bott. Yet just as Porsche introduced Bott's all-wheel drive 964 C4, Branitzki retired, in March 1990. Arno Bohn, a man from the computer industry, replaced him, but after battling the economy, exchange rates, business, and the Supervisory Board, he left on September 30, 1992.

Porsche's third CEO during a single model run stepped into the spotlight

after much hard work backstage. Production engineer Wendelin Wiedeking had supervised massive construction projects, completing them ahead of schedule and under budget. Studying the company's business practices, he had confronted Bott over development costs. He left Porsche in 1988, frustrated by front office and Supervisory Board indifference to his alarms. But Porsche invited him back to the company in 1991 as board member for production. In late 1992, when he assumed his new role, he pledged his own savings account as collateral to establish additional lines of credit for Porsche. In exchange, his contract provided him a bonus if Porsche turned a profit. Few on Porsche's board believed that would happen any time soon. They all believed Wiedeking had even more work cut out for him than any of his predecessors.

▲ Porsche used the 911 Turbo body for these RS and RSR race cars. The street version shown here weighed 1,210 kilograms (2,668 pounds) and developed 300 horsepower at 6,000 rpm. They sold for 225,000DM (roughly $140,600 at the time) though not in the United States.

▲ (Top) The road-going Carrera 3.8 RS, with plaid-upholstered racing seats, a full roll cage, and 300 horsepower with little insulation was barely street civilized. Porsche quoted acceleration from 0 to 100 kilometers per hour in 4.9 seconds. They were available only on special order from Porsche's racing department.

▲ (Bottom) The M64/04 engine used 102mm (4.02-inch) bore and 76.4mm (3.00-inch) stroke for 3,746 cc (228,5-cubic-inch) displacement. For Le Mans, these cars ran with 350 horsepower at 6,900 rpm.

YEAR	1993
DESIGNATION	**911 Carrera RS and RSR 3.8**
SPECIFICATIONS	
MODEL AVAILABILITY	Coupe
WHEELBASE	2272mm/89.4 inches
LENGTH	4275mm/168.3 inches
WIDTH	1775mm/69.9 inches
HEIGHT	1270mm/50.0 inches
WEIGHT	1210kg/2420 pounds (1200kg/2640 pounds RSR)
BASE PRICE	$140,602 ($166,700 RSR)
TRACK FRONT	1440mm/56.7 inches
TRACK REAR	1530mm/60.2 inches (1535mm/60.4 inches RSR)
WHEELS FRONT	9.0Jx18
WHEELS REAR	11.0Jx18
TIRES FRONT	235/40ZR18
TIRES REAR	285/35ZR18
CONSTRUCTION	Unitized welded steel
SUSPENSION FRONT	Independent, lower wishbones, MacPherson struts w/coil springs, gas-filled double-action shock absorbers, anti roll bar
SUSPENSION REAR	Independent, MacPherson struts w/coil springs, gas-filled double-action shock absorbers, anti roll bar
BRAKES	Ventilated, drilled discs, 4-piston aluminum calipers
ENGINE TYPE	Horizontally opposed DOHC six-cylinder Typ M64/04
ENGINE DISPLACEMENT	3746cc/228.6CID
BORE AND STROKE	102x76.4mm/4.02x3.00 inches
HORSEPOWER	300@6500rpm (325@6900rpm RSR)
TORQUE	266lb-ft@5250rpm
COMPRESSION	11.3:1 (11.4:1 RSR)
FUEL DELIVERY	Bosch DME with sequential injection
FINAL DRIVE AXLE RATIO	3.44:1
TOP SPEED	168mph (165mpg RSR)
PRODUCTION	55 (50 RSR)

1994–1997 **911 CARRERA**
1995–1997 **911 GT2 AND GT2 EVO**
1996–1997 **911 GT1**
1997–1998 **911 TURBO S AND CARRERA S**

CHAPTER 6

THE FOURTH GENERATION 1994–1998

1994-1997 911 CARRERA

Ulrich Bez and Harm Lagaay arrived at Weissach in 1989 from BMW. Lagaay pulled designer Tony Hatter off other projects and started him sketching concepts for the next 911. Peter Falk, the competition director, suggested that agility had been engineered out of the 964. He felt its appearance reflected that.

Hatter labored to make this new Porsche very much a 911. He created a significantly changed 911. It had wider, flatter front and rear fenders, which began with raked-back elliptical headlamps and subsided into the trailing edge of the roofline, which then angled downward into the rear bumper. The body provided a slightly larger front trunk and room for the mechanical, safety, and comfort features that buyers had come to expect at that performance and price level.

Performance in all its connotations received prime consideration. Throughout the Weissach campus, engineers strived to address complaints about 964 handling. There was rear suspension noise on both C2 and C4 models. Bez and Paul Hensler intended to provide buyers with more horsepower, but they questioned if that was possible without reinventing the 911 engine.

They reviewed the 965 proposals for water-cooled six- and eight-cylinder engines. The V-8 they ran in the Indianapolis Typ 2708 for Quaker State and Foster's Lager offered possibilities. They discussed another of Bez's interests: a four-door sedan designated the Typ 989. Engineers produced concept drawings and calculations for a 2.5-liter (152.5-

◀ **1996 Ruf Turbo R Coupe**
The reworked engine developed 490 horsepower at 5,500 rpm. Cars weighed 1,491 kilograms (3,280 pounds) and sold for 298,000DM (roughly $198,700) at the factory.

cubic-inch) V-6 that developed 220 DIN horsepower and a turbocharged 3.3-liter (201.3-cubic-inch) V-8 with as much as 408 horsepower. The calculation that killed each of these proposals was cost, so engineering returned to the 3.6-liter flat six in the 964.

Engineer Herbert Ampferer took on the engine. One objective was to eliminate the torsional vibration damper used on the crankshaft to keep the engine running smoothly. By switching to lighter pistons and connecting rods and by making the crankshaft itself stiffer, he succeeded, and from there he and his staff incorporated self-adjusting hydraulically operated valves in the heads. This reduced valvetrain weight and eliminated a service item. When the new engine entered production, it developed 272 DIN horsepower at 6,100 rpm and 243 lb-ft of torque at 5,000. It was quieter and smoother than any 911 engine before it.

With several goals in mind, engineers added a sixth gear to the G50 transmission. This helped the new car meet or exceed U.S. and rest-of-the-world fuel economy standards and reduced engine noise. Although first gear was lower, sixth allowed a top speed of 270 kilometers (169 miles) per hour at 6,700 rpm. At the same time, they improved the four-speed Tiptronic for those who preferred to let Porsche shift gears for them.

Body engineers glued the windshield into the frame rather than letting it float as it had in the past. This increased torsional stiffness and improved handling. Engineers racing against the calendar and budget restrictions to complete the 964 had to attach the rear suspension directly to the car body. This transferred road and suspension noise and movements directly into the interior. The 965 Turbo design introduced parallel

▲ (Top) **1995 Carrera Cabriolet**
The new interior provided drivers a slim airbag-equipped steering wheel as well as new seats and door panels. The windblocker behind the seats was integrated and deployed itself on lowering the roof.

▲ (Bottom) The reworked and heavily revised 3,600cc (219.6-cubic-inch) flat six Typ M 64/05 developed 272 horsepower at 6,100 rpm. A six-speed manual gearbox was standard with American, Austrian, and Swiss buyers getting taller gears from second up to reduce engine noise and emissions and improve fuel economy.

▲ This not only was a new car, it was a distinctively new 911 shape. The characteristic stovepipe front fenders flattened and grew wider, which raised the front deck lid 40mm (1.6 inches) and increased luggage capacity.

wishbones; Fritz Bezner and project manager Bernd Kahnau hoped to adopt this configuration for the new car. The board judged it too costly, but Bez convinced them that the new car needed a better rear suspension if Porsche was to improve on the 964. He had championed the Typ 989 sedan project, and he intended to make it a technological marvel incorporating all-wheel *steering*, among other innovations. Bezner designed a rear suspension system with a lower wishbone and a wishbonelike pair of converging links on top that would accommodate four wheels turning. When Porsche chose not to pursue that expensive technology further, engineers adopted the 928's revolutionary "Weissach" rear axle geometry. This multilink rear suspension countered the transitory effects that induced oversteer or caused rear axle squat on acceleration or lift on hard braking. Bezner's components arrived at Zuffenhausen already mounted on a subframe by the outside supplier. Porsche called it the LSA, for "lightweight," "stable," and "agile," in partial acknowledgment of Peter Falk's attention-getting memorandum.

Where Helmuth Bott's intent with the 964 C4 had been all-wheel traction, Bez and his eventual successor, Horst Marchart, pursued superior handling. Engineers inserted a viscous coupling at the front transaxle to connect the torque tube from the engine to a much smaller (and exposed) differential that split power to each front wheel. This

◀ Designers and engineers reworked the convertible top for the first time since introduction in 1983. Interlocks required drivers to apply the parking brake while raising or lowering the cloth top.

▼ Porsche introduced the next-generation Typ 933 version of the 911 in late 1993 at the Frankfurt International Auto Show. The Carrera 4 coupe and cabriolet arrived for model year 1995. The coupe sold for 134,3400DM in Germany, $69,100 in the United States.

▲ The Cabriolet arrived in the lineup in March 1994. It sold for 142,620DM in Germany, $73,000 in the United States. It accelerated from 0 to 100 kilometers per hour in 5.6 seconds.

assembly weighed 50 kilograms (110 pounds), accounting for half the 964's collection of computers and planetary gears. The lighter weight in front brought the car closer to their goal of agility.

Ulrich Bez pushed Porsche's engineers and designers to make the car look appropriate to its legacy and less expensive and less time-consuming to manufacture. He wanted this new car, the Typ 993, completed in a very short time. He didn't tolerate delays. Weissach staffs knew this character trait, and the backlash it caused reverberated to the Supervisory Board. In the interest of peace, the board chose to not renew Bez's contract. He left in the fall of 1993 but not before he drove a 993 prototype.

Wendelin Wiedeking, who had been back at Porsche for several months by this time, dedicated his efforts to reducing costs and streamlining procedures. He studied Toyota Motor Company's legendary efficiency in purchasing and manufacturing and its adherence to a Japanese philosophy known as *kaizen*, continuous improvement. Two former Toyota executives had founded Shin-Gijutsu, a consultancy dedicated to introducing and embracing new technology, and Wiedeking coaxed the founders to visit Zuffenhausen. What they saw there did not impress them. Viewing parts shelves standing 3 meters (10 feet) tall in the assembly plant, one of them shouted, "Bring us to the factory! This is a warehouse." Wiedeking took a power saw to the shelves. Soon afterward, he adopted "just-in-time" delivery practices that reduced assembly-floor parts inventories from 28 days on hand to 30

▲ Porsche introduced its VarioRam induction system starting with 1996 model year for all 933 Carrera and Carrera 4 versions of the M 64/21 engine. VarioRam debuted on European-only 993 Carrera RS models in 1995. The 3.6-liter flat six developed 285 horsepower at 6,100 rpm.

minutes. Wiedeking forced companies to reexamine their techniques or lose business to competitors who adapted more quickly. Suppliers became subassemblers who, for example, delivered rear shock absorbers mounted in entire rear suspension subframes minutes before assemblers attached them to unibodies.

While Porsche worked toward the 993 launch at the Frankfurt show in late 1993, Wiedeking put the company through the biggest makeover in its history. It was an enormous leap of faith. Before leaving, Ulrich Bez and Chairman Arno Bohn administered a DM 500 million ($310 million) new car program. Yet sales revenues through 1992–1993 had slipped to DM 1.9 billion ($1.2 billion at the time). Zuffenhausen manufactured only 8,341 964 rear- and all-wheel drive and Turbo models. Worse, only 1,188 968s and 119 928S4 models left the plant. Accountants totaled up the year and had to buy red pencils; Porsche set a record it had never sought, losing DM 253 million, or $160 million.

The company assembled 2,374 Typ 993 coupes and another 22 pilot pro-

▲ The cabriolet weighed 1,370 kilograms, 314 pounds. It accelerated from 0 to 100 kilometers per hour in 5.4 seconds with the manual transmission. It sold for 150,800DM in Germany and roughly $78,350 in the United States.

duction cabriolets before the Christmas/New Year holiday break. Named the 911 Carrera in its rear-wheel-drive platform, this first model remained in Europe. Impatient American buyers and journalists received their cars—both coupes and open cars—in April 1995.

"The cabrio was tricky," Tony Hatter said, explaining his work on the top and roofline of the open car. Porsche's first 911 cabriolet had been the 1983 SC. It carried over the same roof through the 964. "I never liked the look of the early cabriolets," he continued. "The classical Nine-Eleven shape is the coupe. With the Nine-Nine-Three, we tried to get some form into the roof. It was the first time, I think, that we tackled the roof."

The 993 Carrera 4 appeared, substantially improved over its 964 predecessor, in midyear as a 1995 model. Customers craving a turbocharger made do with 964 carryover versions until the new 3.6-liter model appeared when the C4 arrived. This was no coincidence; the new Turbo appeared with four-wheel drive and a new six-speed gearbox.

The 993 was a tremendous success. Zuffenhausen assembled 7,074 cabriolets, 7,865 coupes, and 100 new 305-horsepower 993 Carrera Cup cars to update the ongoing series throughout Europe and the United States. Magazine reviewers were pleased, judging in print that the 993 was what they had hoped the 964 would be. The situation was better for everyone; because of Wiedeking's watchfulness and Bez's bullying, Porsche introduced the new cars at a base price $5,000 less than the final 964. By the end of 1994 fiscal year, accountants saw totals in black numerals again.

YEAR	1994-1997
DESIGNATION	911 Carrera
SPECIFICATIONS	
MODEL AVAILABILITY	Coupe, Cabriolet
WHEELBASE	2272mm/89.4 inches
LENGTH	4245mm/167.1 inches
WIDTH	1735mm/68.3 inches
HEIGHT	1300mm/51.2 inches
WEIGHT	1370kg/3014 pounds
BASE PRICE	$69,100 coupe - $73,000 cabriolet
TRACK FRONT	1405mm/55.3 inches
TRACK REAR	1444mm/56.9 inches
WHEELS FRONT	7.0Jx16
WHEELS REAR	9.0Jx16
TIRES FRONT	205/55ZR16
TIRES REAR	245/45ZR16
CONSTRUCTION	Unitized welded steel
SUSPENSION FRONT	Independent, light-alloy lower wishbones, MacPherson struts w/ coil springs, gas-filled double-action shock absorbers, anti roll bar
SUSPENSION REAR	Independent, light-allow multi-wishbone, progressive coil springs, gas-filled double-action shock absorbers, anti roll bar
BRAKES	Ventilated, drilled discs, 4-piston aluminum calipers
ENGINE TYPE	Horizontally opposed DOHC six-cylinder Typ M64/05; Typ M64/07 for US 1994-1995; Horizontally opposed DOHC six-cylinder Typ M64/21; Typ M64/23 for US 1996-1997
ENGINE DISPLACEMENT	3600cc/219.7CID
BORE AND STROKE	100x76.4mm/3.94x3.00 inches
HORSEPOWER	272@6100rpm (M64/05 1994-1995)
	285@6100rpm (M64/21 1997-1998)
TORQUE	243lb-ft@5000rpm (M64/05 1994-1995); 251lb-ft@5250rpm (M64/21 1996-1997)
COMPRESSION	11.3:1 (M64/05) 11.5:1 (M64/21)
FUEL DELIVERY	Bosch DME with sequential injection
FINAL DRIVE AXLE RATIO	3.44:1
TOP SPEED	167mph (M64/05) 171mph (M64/21)
PRODUCTION	14,541 coupes; 7,730 cabriolets in 1994, 1995; 8,586 coupes; 7,769 cabriolets in 1996, 1997

▲ In addressing customer desire to participate in worldwide endurance racing events, Porsche developed a series of cars for the category GT2, more closely based on production cars. These cars, according to co-developer Jürgen Barth, were a cross between Carrera RS models and the 911 Turbo.

1995-1997 911 GT2 AND GT2 EVO

As the economy ground to a halt during the early 1990s, the FIA watched entries at Le Mans drop to just 28 cars in 1992; prototype racing had become too expensive. It ended the Group C series. Their events, especially Gran Turismo—GT-class competition for production-based closed cars—continued drawing entries and audiences. However, many countries had their own rules for GT cars, often slightly different from each other. This made it a corporate gamble to manufacture a car for one series or another. Barth raised the issue with two colleagues, Patrick Peter, who organized the Tour de France for automobiles, and Stéfane Ratel, owner of Venturi, a company with serious racing ambitions. Together the three men launched the BPR series, using the first initials of their last names as the acronym. As competitors themselves, they created rules that made sense to other racers.

Porsche developed cars to meet BPR criteria, starting with the rear-drive 1993 Carrera RS 3.8. Engineers enlarged cylinder bore from 100mm to 102 but kept stroke at 76.4 to achieve this new displacement of 3,746cc (228.5 cubic inches). Using Bosch's Motronic 2.10 system, the engines developed 300 DIN horsepower at 6,500 rpm. Weissach mounted aluminum doors and front deck lids, as well as a fiberglass rear deck lid and spoiler. Underneath, an adjustable anti-roll bar allowed competitors to tune the chassis to individual circuits. Porsche charged DM 225,000 for the cars ($140,625 at the time), and buyers in Europe could option them with radios and air bags for road use or a roll cage and onboard fire extinguishing system for competition. This homologation version helped legalize the RSR 3.8, the real race car welcomed in BPRs GT3 and GT4 events. This all-out machine sold for DM 270,000 (roughly $162,650) and offered center-lock wheels, built-in pneumatic jacks, and supplemental brake cooling. Porsche conservatively rated the engine at 325 DIN horsepower at 6,900 rpm. Throughout 1993 and 1994, the RSRs justified their prices and proved their reputation with overall victories in Spain, Belgium, and Japan and a class win at Le Mans in 1993. The cars routinely claimed class victories throughout 1994.

As the BPR series grew in popularity among entrants and spectators, Barth and his partners added a new series category, GT2. Rules called for manufacturers to assemble 25 cars they intended to enter each year. Barth, Roland Kussmaul, and others within the Customer Sports Department at Weissach developed, introduced, and quickly sold 45 993 GT2s for the 1994 season. (Another 43 followed in 1995, and Weissach sold 14 more in 1996. Porsche equipped 21 of these cars for road use, starting in 1995.)

The GT2 took inspiration from the 911S LM, carrying over its aluminum door panels and front deck lid and its paper-thin side and rear window glass, and eliminating anything not needed for racing. Racing versions weighed 1,112 kilograms (2,447 pounds), while road models were 1,295 kilograms (2,850 pounds). Road cars delivered 430 DIN horsepower through modifications to the 3.6-liter 911 Turbo engine. Porsche sold the car for DM 276,000 ($170,370 in 1994); racers paid a base price of DM 248,500 ($153,395) and got 480 horsepower, but fully optioned racers, with onboard jacks and fire systems, went for as much as DM 335,000 (roughly $223,300) in 1996. Customer Sports built both road and racing models on rear-wheel-drive platforms only, flying in the face of a series production philosophy putting all-wheel drive underneath customer cars with more than 400 horsepower. The GT2 was a practical decision; BPR races had no classes for all-wheel-drive vehicles.

Porsche's BPR series GT3 and GT2 victories, while not quite so ubiquitous or overwhelming as those of its 935s 15 years before, very effectively sold Porsche competition cars to racers and series production models to enthusi-

▲ (Top) Porsche had to manufacture 100 road versions of the car to gain homologation as a racer. This particular car came second in class in the 12 Hours of Sebring in 1995 and won its class in 1996.

▲ (Bottom) With boost turned all the way up and minimal intake restrictor plates in place, the Typ M64/81 engine with 3,600cc displacement (219.6 cubic inches) developed as much as 550 horsepower at 6,000 rpm.

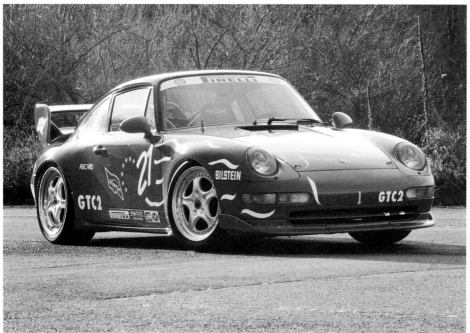

▲ Porsche assembled 43 of these racers. They weighed 1,150 kilograms, 2,535 pounds. Bolted-on fender flares added 40mm (1.6 inches) to the cars width at front and 30mm (1.2 inches) at the rear.

◀ Project manager Roland Kussmaul began developing the 993 Supercup cars as early as May 1993 to be ready for the 1994 season. Internally it was nicknamed the "Cup 3.8." It rode 70mm (2.76 inches) lower than regular production coupes.

▶ For the U.S. Supercup series, this car served as the pace car for 1995 events. It was tuned and developed at an equal level to the racers with stark interior and steel-tube roll cage. Cars accelerated from 0 to 100 kilometers per hour in 4.7 seconds.

asts. At the same time the company introduced the GT2, the Motorsports Department offered a lightened Carrera RS. It removed all insulation, replaced the front hood with an aluminum panel that was 7.5 kilograms (16 pounds) lighter, and substituted side and rear window glass, saving another 5 kilograms (11 pounds). With the same 102mm bore available to Exclusiv customers, the RS setup provided buyers with 300 DIN horsepower in a car that weighed 100 kilograms (220 pounds) less than the standard 993 Carrera 2 coupe. Its most significant innovation was a new engine induction system incorporating variable length intake tubes. This VarioRam system used vacuum-operated sliders to alter intake tube length by providing two intake tubes for each cylinder. Porsche tuned the longer one for maximum midrange torque. At about 4,400 rpm, the shorter one fed more fuel and air for horsepower and torque at higher engine speeds. Then, at 5,800 rpm, the system opened a large cross tube to provide optimal fuel-air mix for the highest engine ranges. The VarioRam not only smoothed out delivery of horsepower and torque, it also added to the driver's enjoyment when engine tones changed as each induction path came into use. The new system was unique to the 1995 RS and its optional Club Sport variation.

The 993 Turbo coupe arrived in spring 1995 at the Geneva Auto Show. The 959 had set a philosophical precedent for Porsche. Management decided that, with exceptions for racing homologations, any series production model providing more than 400 horsepower would have all-wheel drive. The 408-horsepower, 3.6-liter 993 Turbo, putting power to the front wheels as well as through the adapted Weissach rear axle suspension, delivered thrilling performance. Enthusiast magazines regularly recorded acceleration from 0 to 100 kilometers per hour in 4.4 seconds, and the cars topped out at 290 kilometers (180 miles) per hour. Porsche manufactured 2,457 through 1995, following the nine early production pilot cars assembled just before Christmas 1994. Porsche Exclusiv lived up to its name with two other limited edition projects, according to Porsche historian Marc Bongers. Throughout 1995, Exclusiv converted 14 993 Cabriolets into turbocharged open cars by using the 3.6-liter Turbo engine, five-speed manual gearbox, and brakes from the 964 models. In addition, the responsive customer service organization put together a single 993 Speedster for Ferdinand Alexander Porsche.

For 1996 T Program cars, Weissach introduced the VarioRam across the entire normally aspirated lineup, providing them 285 horsepower at 6,100 rpm in street tune. For year-round cruising and touring drivers, Porsche revealed its new Targa. Instead of a fully removable roof panel, the car featured a large glass plate that slid down inside the rear window to open much of the cabin to the skies. Harm Lagaay had conceived this treatment in 1977 as one of his concepts when he designed the 924. The 993 Targa roof fixtures adhered to Wendelin Wiedeking's requirement that outside suppliers deliver intact substructures ready for installation. Zuffenhausen assemblers mounted the roof system onto reinforced cabriolet bodies and welded side rails and front and rear mounts into place.

Porsche achieved another production milestone on July 15, 1996, when its millionth car rolled off the line. The company donated the coupe, fitted with the four-speed Tiptronic S transmission, to the Baden-Württemberg autobahn police in the company's home district.

Porsche's 993 Turbo body, with its larger brakes and wheels and tighter suspension, inspired product planners as it had with the 964. For 1996 they mated the Turbo body and running gear

▲ Intended for Le Mans, the GT2 Evolution ran with Porsche's 3,600cc M64/83 engine that developed 600 horsepower at 7,000 rpm. The cars sold for 570,000DM, roughly $382,000 at the time. *Photograph © 2011 Dave Wendt*

to the normally aspirated 3.6-liter engine and created the Carrera 4S. Porsche offered these cars with the manual transmission only. (For U.S. customers, Zuffenhausen fitted softer springs and shocks to improve ride quality over potholes.) V Program 1997 models saw the addition of the Turbo body look to rear-drive platforms (available with either the six-speed manual or four-speed Tiptronic S transmission) named Carrera S.

To move the road-going benchmark a bit further out, Zuffenhausen's Exclusiv program introduced the 430-horsepower Turbo S for 1997. Porsche claimed the cars reached 300 kilometers (188 miles) per hour and sold them for DM 235,000 (then $130,000). When the model reached U.S. markets in 1998, exchange rates reversed the advantages of earlier times, and Americans paid $155,000 for the S, whose production Porsche limited to just 199 units.

Alois Ruf took each Porsche performance introduction and raised the ante, starting in 1996 with a 993 Turbo-based R coupe developing 490 DIN horsepower. He quoted a top speed of 329 kilometers (204 miles) per hour from the DM 298,000 (about $184,000) coupe. Two years later he introduced his CTR II (narrow body) and CTR II Sport (Turbo body) coupes. These cars he characterized as the "modern successors" to his Yellow Bird. A full car-width rear deck lid supported a hollow rear spoiler that directed air to turbocharger intercoolers. Boasting 520 and 580 brake horsepower respectively, the CTR accelerated to 100 kilometers per hour in 3.6 seconds and topped out at 340 kilometers (213 miles) per hour. Ruf sold them for DM 425,000 (roughly $283,000) at the time. All-wheel drive added another DM 20,700 ($13,800) to the price.

Weissach engineers stretched another limit for 1998, fitting the GT2 flat six with engine management electronics derived from Porsche's Formula One developments with Techniques-Avant Garde (TAG). This increased engine output to 450 DIN horsepower at 6,000 rpm and 430 lb-ft of torque at 4,500. Porsche assembled just 25 of these cars; buyers had the option of air bags, electric window lifts, and air conditioning on the rear-drive platform.

▼ With an eye toward outright victory at Le Mans and in BPR events, Porsche created the 911 GT1 cars. To take advantage of everything the rules permitted, engineers mounted the engine ahead of the rear wheels that lengthened the wheelbase from stock 2,272mm (89.5 inches) to 2,500mm (98.4 inches).

▼ The 3.8-liter M64/70 engine came from the Carrera 3.8 RS model, tuned in this case to develop 310 horsepower at 6,100 rpm. It ran on super-unleaded fuel and competed with its oxygen sensor and catalytic converter in place.

▲ Stiffer springs and a new rear multi-link suspension greatly improved the handling of the Supercup cars. They weighed 1,100 kilograms, 2,420 pounds, complete with air-jack system.

▶ The GT1 grew from stock 993 dimension in all directions. Compared with the production car overall length of 4,245mm (167.1 inches), the GT1 stretched 4,683mm (184.4 inches). At 1,946mm wide, it was 211mm wider (68.3 inches to 76.6 inches).

YEAR	1995-1997
DESIGNATION	911 GT2 and 911 GT2 Evo

SPECIFICATIONS

MODEL AVAILABILITY	Coupe Racing	WHEELS REAR	11.0Jx18 (both)	ENGINE DISPLACEMENT	3600cc/219.7CID
	Coupe Road	TIRES FRONT	265/645-18 (race) – 235/40ZR18 (road)	BORE AND STROKE	100x76.4mm/3.94x3.00 inches
WHEELBASE	2272mm/89.4 inches			HORSEPOWER	450@5700rpm (race)
LENGTH	4245mm/167.1 inches	TIRES REAR	305/645-18 (race) – 285/35ZR18 (road)		430@5750rpm (road)
WIDTH	1935mm/76.2 inches (race)				600@7000rpm (EVO)
	1855mm/73.0 inches (road)	CONSTRUCTION	Unitized welded steel	TORQUE	470lb-ft@5000rpm (race) —
HEIGHT	1300mm/51.2 inches (race)	SUSPENSION FRONT	Independent, light-alloy lower wishbones, MacPherson struts w/coil springs, gas-filled double-action shock absorbers, anti roll bar		398lb-ft@4500rpm (road)
	1270mm/50.0 inches (road)				479lb-ft@ 4,000rpm (EVO)
WEIGHT	1100kg/2420 pounds (race)			COMPRESSION	8.0:1
	1295kg/2849 pounds (road)			FUEL DELIVERY	Bosch Motronic TAGTronic fuel injection, twin turbochargers, intercoolers
BASE PRICE	$238,800 (race) — $193,007 (road)				
		SUSPENSION REAR	Independent, light-allow multi-wishbone, progressive coil springs, gas-filled double-action shock absorbers, anti roll bar (adjustable in car)		
	$382,500 (EVO)			FINAL DRIVE AXLE RATIO	Varies by circuit (race) – 3.44:1 (road)
TRACK FRONT	1454mm/57.2 inches (race)				
	1475mm/58.1 inches (road)			TOP SPEED	Varies by final drive (race) – 167mph (road)
TRACK REAR	1540mm/60.6 inches (race)	BRAKES	Ventilated, drilled discs, 4-piston aluminum calipers		
	1550mm/61.0 inches (road)			PRODUCTION	43 in 1994 (race) — 172 (road, between 1995 and 1997)
WHEELS FRONT	10.0Jx18 (race) – 9.0Jx18 (road)	ENGINE TYPE	Horizontally opposed DOHC six-cylinder Typ M64/81; Typ M64/83		

1996-1997 911 GT1

Porsche held a solid lock on GT2 and GT3 class wins, but overall victories in BPR's endurance series went to premier category entrants racing in GT1. During the Peter Schutz years, engineers had grown accustomed to winning, and Norbert Singer knew the 911 GT2 needed better aerodynamics, greater downforce, and more horsepower to grab the lead. Rules required that the car have a flat bottom from its nose to the rear axle. Behind that point, Singer could create one or more venturi to hold the tail down. Following Porsche's tradition of mounting the engine behind the rear axle severely limited the space available for these cavities. Singer and Herbert Ampferer reversed the engine, installing it ahead of the rear axle. BPR rules allowed longer wheelbases for GT1 entries, so they lengthened the GT2 from 2,270mm (89.4 inches) to 2,499mm (98.4 inches) for the new car.

BPR also required GT1 manufacturers to offer road versions of the car as well, though it became clear that a single car or two might satisfy the regulators. Starting in 1995, engineer Horst Reitter worked with Singer to establish package specifications, including steering, suspension, engine, drivetrain, cooling, and cockpit. Reitter incorporated the production 993 front end because it already had passed German and U.S. crash tests, necessary for the single road car's homologation. This allowed him to install the 993 instrument panel as well. One requirement from Porsche Supervisory Board was more challenging than the BPR specifications: The car had to be "identifiable as a 911 at first glance."

That job fell to 993 stylist Tony Hatter. At first Norbert Singer was not thrilled working with the Design Department. Singer, a man with decades of hands-on experience, was skeptical,

▼ While the production 993 stood 1,300mm tall (51.2 inches), the GT1 hunkered down to 1,173mm (46.2 inches). It weighed just about 1,000 kilograms, 2,200 pounds. Porsche's GT1s finished second and third overall at Le Mans in 1996.

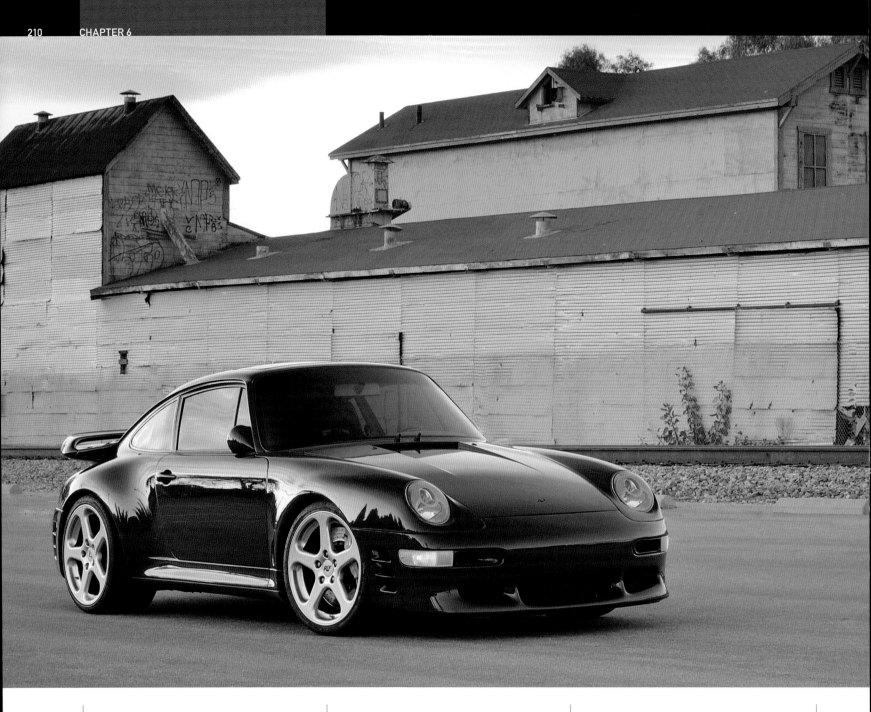

▲ Alois Ruf developed his Turbo R based on Porsche's 993 Turbo model, fitting the car with firmer springs and larger anti-roll bars and altering shock absorber characteristics.

and Hatter had to do much of his work when Singer left the room. Then he learned that styling had computer technology that let them build cars digitally. Singer and Hatter created shapes and milled them into three-dimensional wind tunnel models very quickly. Hatter lengthened the 993 and widened it to accommodate racing tires. Together they invented Porsche's first mid-engine 911. The board approved it in late July 1995 for the 1996 racing season.

It required substantial effort to fit the production nose over Reitter's racing suspension. Herbert Ampferer took on power train development, revising the successful 962 engines with their 95mm bore and 74.4mm stroke for a total displacement of 3,164cc (193.0 cubic inches). He water-cooled the entire engine. For the single road-going example for homologation, Weissach installed a standard 3.3-liter (201.3-cubic-inch) Carrera engine tuned to produce 300 brake horsepower. But development continued on the racing models. Jürgen Barth drove the first prototype in mid-March, barely a month before the Le Mans April trials.

Singer and his staff tested the car extensively, running up nearly 2,000 kilometers (1,200 miles) in a single five-day test. Porsche entered chassis 001 and 002 at Le Mans in June 1996. When the race ended, 002 had taken first in the GT1 class, second overall behind Reinhold Joest driving a Porsche-powered prototype.

Porsche fielded dozens of orders for the road car, which slowed only when Barth informed potential customers they were getting a 300-horsepower street engine, not the 600-horsepower Le Mans version. Meanwhile, privateers

▲ Ruf mounted 18-inch alloy wheels. Acceleration from 0 to 100 kilometers per hour took 3.6 seconds. With appropriate gear ratios, top speed was 329 kilometers (204 miles) per hour.

▶ Small touches were everywhere in the interiors. Ruf instruments provided green markings. His steering wheel contained an airbag, and he re-contoured the sports seats.

▼ (Left) The water-cooled Typ M96/80 engine, with 95mm (3.74-inch) bore and 74.4mm (2.93-inch) stroke, displaced 3,163cc (192.9 cubic inches). It developed 600 horsepower at 7,200 rpm.

▼ (Right) Engineer Hans Reitter conceived of using the lower section of the production 933 front end. Rules required for manufacturers to sell road-going versions of these racers meaning they had to survive a crash test. That left a race car interior that looked familiar to most Porsche drivers.

placed requests for the competition model. The racers weighed 1,058 kilograms (2,328 pounds), and with long gears for Le Mans, they reached 375 kilometers (235 miles) per hour. Porsche had created yet another out-of-the-box success. At Brands Hatch in England, Spa in Belgium, and Zhuhai in China, car 001 or 002 took first place. Then Weissach went to work filling nearly 30 solid orders for racing and road versions of the car known as GT1/96. Porsche set a price of DM 1,550,000 ($1 million) for the racers and DM 1,400,000 (roughly $890,000) for road cars, which ultimately ran 544-horsepower engines, not the lesser homologation motors. Paul Frère, writing in *Road & Track* magazine, put the road version in context: Private individuals will buy the GT1, he wrote, "for the pleasure of driving the closest possible thing to a full-blooded, very high performance racing car." This taught Porsche another marketing lesson.

With competition increasing in the rarified GT1 category, Porsche decided to support its 30-some customers with updates and a new version, the GT1/97. Norbert Singer made enough changes that Porsche had to create a 1997 GT1 road car as well, picking up styling cues from the next generation 911. Sanctioning body BPR yielded authority to the FIA, which offered to run an 11-race series for 1997.

It was a less successful season for Porsche. With losses to Mercedes-Benz and McLaren, Singer, Barth, and their customers knew victory in 1998 required an all-new car. Using carbon fiber for the chassis saved 90 kilograms (200 pounds), which everyone felt was necessary. Wiedeking agreed, and he sanctioned GT1/98. But a new 911 was coming as well, with a new look and new engine, and any Porsche race car that had "to be identifiable as a 911 at first glance" had to reflect this new appearance.

▲ **1998 Ruf Turbo R Cabriolet**
Alois Ruf was concerned about putting such performance potential into an open car without his typical roll cage. But this longtime customer convinced Ruf this car was for sunny day drives along the California coast.

◀ It is a characteristic of Ruf's automobiles, and of those from Porsche as well, that interiors were understated. The performance of the cars made the comments anyone thought necessary.

▲ The cabriolet weighed 1,491 kilograms, 3,280 pounds. Acceleration from 0 to 100 kilometers per hour took 3.6 seconds.

▲ The owner described the color as California Poppy. An integrated rear wind deflector rose at the top descended.

▲ The car rode on 225/40ZR18 tires in front and 285/30ZR18s in the rear. Top speed was reported in excess of 325 kilometers (203 miles) per hour.

YEAR	1996-1997
DESIGNATION	911 GT1
SPECIFICATIONS	
MODEL AVAILABILITY	Coupe
WHEELBASE	2500mm/98.4 inches
LENGTH	4638mm/184.4 inches
WIDTH	1946mm/76.6 inches
HEIGHT	1173mm/46.2 inches
WEIGHT	1000kg/2205 pounds
BASE PRICE	$890,805
TRACK FRONT	1502mm/59.1 inches
TRACK REAR	1588mm/62.5 inches
WHEELS FRONT	11.5Jx18
WHEELS REAR	13.0Jx18
TIRES FRONT	27/68-18
TIRES REAR	30/70-18
CONSTRUCTION	Unitized sheet steel, rear tube frame
SUSPENSION FRONT	Independent, tubular steel upper and lower wishbones, Bilstein shock absorbers with coil springs, anti roll bar
SUSPENSION REAR	Independent, upper and lower A-arms with pushrods, anti roll bar
BRAKES	Ventilated, drilled discs, 8-piston front, 4-piston rear calipers
ENGINE TYPE	Horizontally opposed DOHC six-cylinder Typ M96/80, water-cooled cylinder heads
ENGINE DISPLACEMENT	3163cc/193.0CID
BORE AND STROKE	95x74.4mm/3.74x2.93 inches
HORSEPOWER	544@0700rpm
TORQUE	434lb-ft@6250rpm
COMPRESSION	9.0:1
FUEL DELIVERY	TAG engine management, twin turbochargers, intercooler
FINAL DRIVE AXLE RATIO	3.44:1 (varies)
TOP SPEED	193mph (varies)
PRODUCTION	21

▲ Porsche *Exclusiv* developed its most exclusive Turbo, the S model, for customers startling in late 1997 model year. The bodywork included the Aerokit II, with modified front and rear spoilers and additional front slots for greater brake cooling.

1997-1998 911 TURBO S AND CARRERA S

Such regulations did not affect Porsche's ability to service clients who had not gotten enough of some of the 993's best vehicles. As the company prepared the new model, it offered a final run of narrow-body Targas and wide-body Carrera S coupes on the rear-drive platform. All-wheel-drive versions of the 993 Carrera 4S coupe appeared one last time, as did the 408-brake-horsepower Turbo coupe. Exclusiv offered yet another home-market version—with a Performance Kit utilizing two larger turbochargers, an additional oil cooler, and a reprogrammed engine management computer—providing 450 brake horsepower at 6,000 rpm, at a DM 29,800 (approximately $16,900) premium over the DM 222,500 price ($126,420 at the factory). The car took 4.1 seconds to reach 100 kilometers per hour. Because Porsche often held the best to last, Exclusiv introduced a one-year-only 450-horsepower Turbo S, fitting the body with the full Aerokit II, as well as extensive carbon fiber throughout the interior. Porsche assembled 345 of these special 1,500-kilogram (3,300-pound) cars, which sold through the 1998 model year for DM 307,300 ($174,600). The only 1998 model more potent was the 1,295-kilogram (2,849-pound), 450-horsepower GT2, of which 21 sold for DM 287,500 ($163,352).

These cars represented the end of a 50-year era of air-cooled Porsche cars, and some owners ranked the Typ 993 as the last pure 911. For an even larger number, the next-generation 911, with its new body, interior, engine, and sound, was the first 911 they ever considered buying.

▲ The 3,600cc (219.6-cubic-inch) M64/60S engine developed 450 horsepower at 6,000 rpm in the S. Acceleration from 0 to 100 kilometers per hour took 4.1 seconds. Porsche quoted a top speed of 300 kilometers (186 miles) per hour.

▶ The interior was a blend of leather and carbon fiber on the Turbo S models. Everything that wasn't carbon fiber was leather-covered.

▲ Porsche's Turbo coupe with the Performance Kit developed 430 horsepower from the M64/60R engine at 5,750 rpm. This was good for acceleration from 0 to 100 kilometers in less than 4.5 seconds. Top speed was reported as more than 290 kilometers (180 miles) per hour. *Photograph © 2011 Dave Wendt*

▲ (Top) These were not prototype Porsche Carbon Composite Brakes, merely calipers painted Speed Yellow as part of the Turbo S package. The car rode 15mm (0.6 inches) lower by virtue of the sport suspension tuning. The car sold for 304,650DM and $150,000 in the United States.

▲ (Bottom) Porsche chromium-plated the standard Turbo S instrument bezels and finished the instrument faces in silver paint. The Turbo S logo was embroidered or embossed in several places throughout the car.

YEAR	1997-1998
DESIGNATION	911 Turbo S
SPECIFICATIONS	
MODEL AVAILABILITY	Coupe
WHEELBASE	2272mm/89.4 inches
LENGTH	4245mm/167.1 inches
WIDTH	1795mm/70.7 inches
HEIGHT	1285mm/50.6 inches
WEIGHT	1500kg/3300 pounds
BASE PRICE	$175,086
TRACK FRONT	1411mm/55.6 inches
TRACK REAR	1504mm/59.2 inches
WHEELS FRONT	8.0Jx18
WHEELS REAR	10.0Jx18
TIRES FRONT	225/40ZR18
TIRES REAR	285/30ZR18
CONSTRUCTION	Unitized welded steel
SUSPENSION FRONT	Independent, light-alloy lower wishbones, MacPherson struts w/coil springs, gas-filled double-action shock absorbers, anti roll bar
SUSPENSION REAR	Independent, light-allow multi-wishbone, progressive coil springs, gas-filled double-action shock absorbers, anti roll bar
BRAKES	Ventilated, drilled discs, 4-piston aluminum calipers
ENGINE TYPE	Horizontally opposed DOHC six-cylinder M64/60S
ENGINE DISPLACEMENT	3600cc/219.7CID
BORE AND STROKE	100x76.4mm/3.94x3.00 inches
HORSEPOWER	450@6000rpm
TORQUE	431lb-ft@4500rpm
COMPRESSION	8.0:1
FUEL DELIVERY	Bosch DME with sequential injection, turbochargers, intercooler
FINAL DRIVE AXLE RATIO	3.44:1
TOP SPEED	186mph
PRODUCTION	345

1998-2001 **911 CARRERA**
1999-2005 **911 GT3**
2002-2004 **911 CARRERA AND TURBO**
2003-2005 **911 GT2**
2001-2005 **911 TARGA**
2003-2004 **911 CARRERA 40TH ANNIVERSARY EDITION**

CHAPTER 7

WATER COOLING DEFINES THE FIFTH GENERATION 1998–2005

1998-2001 911 CARRERA

As iconic as the air-cooled engine in the 911 had become, eventually Porsche decided it was the time to introduce water-cooled engines. Noise regulations and emissions restrictions made it necessary. Porsche's ethic of always providing more powerful successor models made it essential.

"Water cooling," Stefan Knirsch explained, "allowed us to get higher performance because of the better cooling of the cylinder head." Knirsch joined Porsche in 1996 when the engines were in preproduction phases. He was hired as troubleshooter at the start of manufacture of the new 911. "The only drawback," he said, "was additional mass, the twenty liters of water and all the parts. You need low temperatures of the components, of the cylinder head and the block, to get a high output and good fuel economy."

The new engine was a radical and complicated departure from what had come before (and been used for nearly 35 years at Porsche). To facilitate assembly, engineers designed the aluminum alloy crankcase and cylinder head assembly to split down the middle. During the foundry operation, Porsche cast in place the aluminum-silicon alloy cylinder liners. A separate shaft powered by chains ran below the crankshaft to drive four separate camshafts, also chain operated. The VarioCam system allowed 25 degrees of adjustment of intake cam timing and opening, depending on engine speed and load. The finished product displaced 3,387cc (206.6 cubic inches) with 96mm bore and 78mm stroke, dimensions referred

◀ The 996's most significant development was its water-cooled flat six M96/01 engine. With bore of 96mm (3.78 inches) and stroke of 78mm (3.07 inches), overall displacement was 3,387cc (206.6 cubic inches).

to as over square or short stroke, to provide smooth running high engine speeds. With 11.3:1 compression, the engine developed 296 DIN horsepower at 6,800 rpm and 258 ft-lb of torque at 4,600 rpm.

Water cooling eliminated cooling fins, leaving room in the heads for four valves. Designers returned to a single spark plug for ignition. Marchardt refused to subject buyers to oil leaks, which plagued the complicated plumbing of dry sump engines and led to complaints and service visits. With water cooling, and because this engine was designed for road use only, there no longer was any need. Liquid cooling improved heating and cooling capabilities as well, even as it blunted the mechanical noises of the engine. To preserve some noise (and driving excitement), engineers revised the VarioRam induction, tuning intake and exhaust manifolds not only for efficient fuel flow but for the sound these processes produced as well.

Porsche and Getrag collaborated on a new six-speed transmission with the capacity to handle much more power down the line. Increasing engine output always was a company target, and Marchardt had participated in the model proliferation that accompanied 3.2 Carrera, 964, and 993 lines. The updated and strengthened Tiptronic S gearbox provided five speeds for the new 911 lineup, designated the 996. Drivers could shift manually either on the center console or using rocker switches on the steering wheel.

Harm Lagaay launched an internal contest to select the designers who would take on the 996 and the entry-level car, the Typ 986. Stylist Pinky Lai's work won him the assignment for the new 911. Marchardt's orders for the car included improving aerodynamics and further reducing rear lift.

◀ The front end of the Typ 996 was nearly identical to the Boxster Typ 986 that debuted a year earlier. While this development saved Porsche millions of Deutschmarks, it led to confusion at first about which car people were seeing.

▲ In profile, the 996 was more distinctive. The wheelbase grew from 2,272mm (89.5 inches) for the 993 to 2,350mm (92.5 inches) for the new car. Engineers quickly filled the space.

"Budget restrictions were extremely tight," Pinky Lai explained.

The 996 platform team had told Lai there was no money to incorporate a retractable rear wing, as Porsche had done on 964 and 993 models. "They had a very tough bean counter," Lai explained. "He showed up every day. 'You can only design a body that will meet the aerodynamic target.'" One day in the wind tunnel, he and the engineers trimmed off one of the rear grille louver panels and remounted it backward. It changed the readings immediately. "We came up with the argument that it wasn't a moving spoiler, it was a moving grille."

The 911 expanded in all directions with the 996. The wheelbase added 3 inches, going from 2,272mm on the 993 to 2,350mm for the new car. Overall length stretched 7.3 inches, growing from 4,245 to 4,430mm (174.4 inches), and width grew from 1,735 to 1,765mm (69.5 inches). Height increased only incrementally, by 5mm to 1,305 (51.4 inches). Despite these gains in every dimension and the addition of water-cooling paraphernalia, curb weight dropped 50 kilograms (110 pounds) to 1,320 kilograms (2,904 pounds). In addition to weight reduction, the diligence designers and engineers showed with the body design reduced the Cd, from 0.34 for the 993 to 0.30 for the new car.

With a goal of making the 996 easier and less expensive to manufacture, design and production engineers

▲ Porsche introduced the 996 Cabriolet in April 1998. The redesigned convertible top went up or down in 20 seconds with the vehicle parked. The car sold for 155,160DM in Germany, $73,000 in the United States.

▶ The coupe grew in all dimensions from the previous 993. But in one important measurement, the number shrunk. The 993 had weighed 1,370 kilograms (3,014 pounds). Even with the addition of water-cooling radiators and tubing, Porsche brought the 996 down to 1,320 kilograms, 2,904 pounds.

▲ The new 3.4-liter engine developed 300 horsepower at 6,800 rpm. Acceleration from 0 to 100 kilometers per hour took 5.2 seconds.

made use wherever possible of lighter-weight and stronger metals and other materials. High-strength steel in structural body panels, assembled in a process called tailored blanks, incorporated careful edge trimming that left only enough metal for welding; this complex process increased torsional stiffness of the new car by 45 percent and bending stiffness by 50 percent. The fully independent front suspension used MacPherson struts, and a five-link configuration supported the rear. This entire system induced slight understeer at cornering limits, which Porsche engineers concluded was a more comfortable characteristic to many drivers.

What's more, this combination, including the shock absorbers and springs, gave the car a more comfortable ride. For those who wanted their sports car to be sporty, an optional suspension lowered ride height 10mm (0.4 inches), and fitted stiffer springs, more responsive shock absorbers, and larger diameter anti-roll bars front and rear. The car's lighter weight, coupled with the 11-horsepower (DIN) increase to 296, brought quicker acceleration (from 0 to 62 miles per hour in 5.2 seconds, compared to 5.6 for the 993) and a higher top speed (278 kilometers [174 miles] per hour, compared to 267 kilometers [167 miles] per hour for the earlier car). At introduction, Porsche charged DM 135,610 for the base coupe with six-speed manual transmission ($65,030 in the United States).

Porsche followed the coupe with the 996 cabriolet in April 1998. This model came with an aluminum removable hardtop with an electrically heated rear window. It weighed 33 kilograms (73 pounds). The electrically operated cloth hood folded up or down in a Z movement in 20 seconds when the vehicle was parked.

Six months after the cabrio appeared, the 996 line expanded in October 1998 when Zuffenhausen began assembling all-wheel-drive Carrera 4 coupes and cabriolets. As with rear-drive C2 and all-wheel drive C4 models in the past, bodies were identical except for specific badging. Porsche's third-generation all-wheel-drive system improved on both predecessors by using an even lighter open driveshaft. Weissach engineers incorporated the viscous clutch in the front differential. The company offered the Tiptronic S five-speed transmission to C4 customers as well as the six-speed manual. A new sophisticated traction control system, Porsche Stability Management (PSM), could slow engine speed and/or apply brake pressure to one or more wheels. PSM incorporated an electronic throttle control, called E-Gas, in which a potentiometer read gas pedal position and transmitted that information to the Bosch Motronic ME 7.2 engine management system. A servomotor reacted to electronic inputs and opened or closed throttle butterflies in the VarioRam induction system. Rear-drive Carreras still operated with a traditional throttle cable.

▲ According to Harm Lagaay, Porsche's head of design, this multi-function headlight fixture resulted from production demands to install front lights in 20 seconds. This incorporated low and main beams, parking and fog lamps, turn signal and headlight washer in one assembly.

▲ The larger exterior dimensions yielded more interior space for driver and passenger. Seats with leather surfaces adjusted the backrest angle electrically with manual fore, aft, and height adjustment.

◀ The removable hardtop nearly reproduced the lines and shapes of the 996 Carrera coupes. Rear suspension on the 996 was a multi-link system with its own subframe that improved handling yet isolated road noise and vibration.

▲ With the hardtop in place, the cabriolet became a comfortable, secure all-weather car. The cloth top remained in place below its metal cover.

▲ Cabriolet buyers also received an aluminum hardtop as standard equipment. It weighed 33 kilograms (77 pounds).

▲ Weissach engineers developed a new front suspension for the 996. The MacPherson strut configuration incorporated "disconnected" longitudinal and transverse links joined by elastic rubber bushings.

▲ **2002 Ruf Turbo Coupe**
Alois Ruf gave customers the choice of their Turbo on the all-wheel-drive platform or strictly a rear-drive version. They also could select a narrow body or the standard wide configuration.

▶ The rear wing with electric adjustments was Ruf's design, as were the outside mirrors. The company quoted a top speed of 330 kilometers (205 miles) per hour. The R Turbo sold for 219,588DM, roughly $198,000.

▲ Ruf's engineers coaxed 520 horsepower at 6,000 rpm out of his 3,600cc (219.6-cubic-inch) water-cooled engine. Acceleration from 0 to 100 kilometers per hour took just 3.7 seconds.

◀ Ruf's instruments used green markings. The sport steering wheel and slim racing-type seats were standard equipment. A six-speed manual transmission was standard or buyers could order the next generation five-speed Tiptronic S.

YEAR	1998-2001
DESIGNATION	**911 Carrera 4 introduced in 1999**
SPECIFICATIONS	
MODEL AVAILABILITY	Coupe, Cabriolet
WHEELBASE	2350mm/92.5 inches
LENGTH	4430mm/174.4 inches
WIDTH	1765mm/69.5 inches
HEIGHT	1305mm/51.3 inches
WEIGHT	1320kg/2904 pounds
BASE PRICE	$65,030 Carrera Coupe
	$74,460 Carrera Cabriolet
	$70,480 Carrera 4 Coupe
	$79,920 Carrera 4 Cabriolet
TRACK FRONT	1455mm/57.3 inches
TRACK REAR	1500mm/59.1 inches
WHEELS FRONT	7.0Jx17
WHEELS REAR	9.0Jx17
TIRES FRONT	205/50ZR17
TIRES REAR	255/40ZR17
CONSTRUCTION	Unitized welded steel
SUSPENSION FRONT	Independent, light-alloy wishbones, MacPherson struts w/coil springs, gas-filled double-tube shock absorbers, anti roll bar
SUSPENSION REAR	Independent, multi-wishbone, progressive coil springs, gas-filled single-tube shock absorbers, anti roll bar
BRAKES	Ventilated, drilled discs, 4-piston aluminum monobloc calipers
ENGINE TYPE	Horizontally opposed water-cooled DOHC six-cylinder Typ M96/01
ENGINE DISPLACEMENT	3387cc/206.7CID
BORE AND STROKE	96x78mm/3.78x3.07 inches
HORSEPOWER	300@6800rpm
TORQUE	258lb-ft@4600rpm
COMPRESSION	11.3:1
FUEL DELIVERY	Bosch DME with sequential injection
FINAL DRIVE AXLE RATIO	3.44:1
TOP SPEED	174mph
PRODUCTION	31,135 Carrera coupes; 25,598 cabriolets;
	12,643 Carrera 4 coupes; 9,411 Carrera 4 cabriolets;
	from 1998 through 2001

1999-2005 911 GT3RS

Porsche debuted its 996 Carrera Cup racers late in the 1998 season, launching the cars in April as part of the Pirelli Supercup series. For the German Carrera Cup series, cars ran the full 1999 season. These were pure race cars, whose ride height Porsche Motorsports had lowered by 60mm (2.36 inches) from standard road-going Carreras. Suspensions were fully tunable, from height-adjustable shock absorbers to variable stiffness front and rear anti-roll bars to wide ranges of camber and toe-in settings. Porsche mounted Pirelli racing slicks—245/645x8 front and 305/645x18 rear—on center-lock wheels, inside which Motorsports fitted massive 330mm brake rotors with four-piston fixed calipers and a modified ABS. The 3.6-liter Mezger motors initially developed 360 DIN horsepower at 7,250 rpm, providing startling performance to the 1,140 kilogram (2,513 pound) Cup coupes. Engineers improved engine output to 370 DIN horsepower at 7,200 rpm for 1999. For 2000 Cup cars went to 305/660x18 rear tires. Porsche produced 81 of the cars in 1999 and 137 in 2000.

To qualify for other FIA competitions, the company introduced a 996 GT3 coupe in May 1999 as a 2000 model year offering. Porsche limited sales to mainland Europe and the United Kingdom. Unlike the standard 996, the GT3 used a water-cooled derivative of the earlier Hans Mezger–designed 3,600cc (219.6-cubic-inch) flat six with dry sump lubrication, accomplished by merging the multivalve, water-cooled heads of the 3.4-liter 996 production series with the bottom end of the engine from the Typ 964 and the 1998 Le Mans–winning GT1. With appropriate tweaks to the VarioRam and the engine management computers, this engine developed 360 DIN horsepower at 7,200 rpm. It coupled to the road with the latest version of the GT2 transmission.

To improve handling for those buyers who really intended to race, Weissach engineers lowered the GT3 suspension by 30mm (1.2 inches). Brake rotors grew to 330mm all around (instead of 318 fronts and 299 rears). Weissach engineers kept the ABS but deleted PSM as well as rear seats, door speakers, and more than half the sound insulation. To offer the car to most of the world, Porsche retained front and side air bags and included electric seat adjustment and electric windows (they weighed less than mechanical lifts and would have required modification to replace them). Porsche offered a radio and air conditioning as no extra cost options.

The GT3, like the 993 GT2 before it, looked very much like a standard production 911 except for the elevated fixed wing on the rear deck. As with previous RS models, Porsche offered both a standard "Comfort" version, as described, and a "Clubsport" option,

(CONTINUED ON PAGE 243)

▼ Many parts were plastic: doors, front fenders, rear deck lid and wing, and the nose. With the gutted interior and ruthless lightening efforts, the competition RS weighed 1,110 kilograms (2,447 pounds).

▲ 2002 996 GT3 RS Coupe
The 3.6-liter water-cooled M96/77 engine developed 435 horsepower at 8,250 rpm. Porsche manufactured 48 of these coupes.

◀ Racing engineer and driver Roland Kussmaul developed these new cars over an intense year of effort. The nosepieces allowed extra cooling inlets for brakes and the engine.

▲ 2004 996 GT3 Coupe
American buyers got their hands on the GT3 model starting in 2004. The 3,600cc 219.6-cubic-inch water-cooled M96/79 engine developed 381 horsepower at 7,400 rpm.

▲ Essentially this was the road-going version of Porsche's popular and successful Porsche Carrera Cup race cars. Unlike the Cup cars, leather covered sport seats, air conditioning, and a CD player were standard equipment.

▲ The fixed rear wing was well suited to track-days participants. Weissach engineers designed the spoiler to offer three different angle settings.

▲ The GT3 accelerated from 0 to 100 kilometers per hour in 4.5 seconds. Porsche quoted a top speed of 306 kilometers (190 miles) per hour. Cars sold for €102,112 in Germany and $99,000 in the United States.

▲ Porsche chose not to export its 1999 and 2000 GT3 models to the United States, due to Weissach decisions to remove driver airbags from the cars to save weight. By 2004, when the next generation cars appeared, America buyers were anxious and enthusiastic customers. *Photograph © 2011 Dave Wendt*

▲ 2004 996 GT3 RS "Street" Coupe

The company needed to assemble at least two hundred of these coupes. Through judicious use of carbon fiber and some plastic in windows, Weissach engineers trimmed 50 kilograms from the already lightened Club Sport models, to achieve 1,360 kilograms, 2,998 pounds.

▲ Racing fans among Porsche's customers benefited from the company's participation in GT events and the need to produce minimum numbers of roadworthy examples to qualify a racer for a series. This GT3 RS was one of many such vehicles.

▲ The 3,600cc (219.6-cubic-inch) M96/79 engine developed 381 horsepower at 7,300 rpm. Acceleration from 0 to 100 kilometers took 4.4 seconds, and Porsche quoted its top speed at 306 kilometers (190 miles) per hour.

▶ Porsche offered the RS only in Carrera White with red or blue script on the doors and rear valence as well as wheels. The company did not export these cars to the United States or Canada.

▲ Other than the obvious roll cage, racing seats with six-point harnesses, and on-board fire system, the interior was hard to differentiate from the base Carrera model with radio, climate control, and electric windows.

(CONTINUED FROM PAGE 234)

which replaced leather seats with race bucket seats and provided a six-point harness for the driver, a roll cage, and a fire system mounted in the passenger footwell. Production figures indicate that Porsche assembled 1,868 of the Comfort and Clubsport editions, both of which it sold at the factory for DM 179,500 ($86,298 at the time). The GT3 accelerated from 0 to 62 miles per hour in 4.8 seconds and went on to a top speed of 302 kilometers (188 miles) per hour.

With the GT3 unavailable to U.S. customers, Alois Ruf stepped into the hole in the market with his RGT based on the 996. Using the Mezger 3.6-liter engine, Ruf's staff retuned the engine control computers, modified all four camshafts, and fitted a low restriction air intake filter and a high-performance exhaust. They derived 385 DIN horsepower from the engine and had the automobile certified by EPA and DOT regulators, as well as those in the rest of the world. Inside the car, Ruf trimmed the interior in leather and integrated a roll cage into the A-, B-, and C-pillars. Acceleration took 4.6 seconds from 0 to 62 miles per hour, and Ruf's testers reached 307 kilometers (191 miles) per hour in the car. He sold them for $135,000 in the United States.

YEAR	2006-2008
DESIGNATION	911 GT3, GT3 RS
SPECIFICATIONS	
MODEL AVAILABILITY	Coupe
WHEELBASE	2360mm/92.9 inches
LENGTH	4445mm/175.0 inches
	4460mm/175.6 inches GT3 RS
WIDTH	1808mm/71.2 inches
	1852mm/72.9 inches GT3 RS
HEIGHT	1310mm/51.6 inches
WEIGHT	1395kg/3069 pounds
	1375kg/3025 pounds GT3 RS
BASE PRICE	Not available
TRACK FRONT	1486mm/58.5 inches
TRACK REAR	1519mm/59.8 inches
WHEELS FRONT	8.5Jx19
WHEELS REAR	12.0Jx19
TIRES FRONT	235/35ZR19
TIRES REAR	305/30ZR19
CONSTRUCTION	Monocoque steel
SUSPENSION FRONT	Independent, wishbones, semi-trailing arms, MacPherson struts w/coil springs, gas-filled double-tube shock absorbers, anti roll bar
SUSPENSION REAR	Independent, multi-wishbone, progressive coil springs, gas-filled single-tube shock absorbers, anti roll bar
BRAKES	Ventilated, drilled discs, four-piston aluminum monobloc calipers
ENGINE TYPE	Horizontally opposed water-cooled DOHC six-cylinder M97/76
ENGINE DISPLACEMENT	3600cc/219.6CID
BORE AND STROKE	100x76.4mm/3.94x3.01 inches
HORSEPOWER	415@7600rpm
TORQUE	299lb-ft@5500rpm
COMPRESSION	12.0:1
FUEL DELIVERY	Bosch DME with sequential injection
FINAL DRIVE AXLE RATIO	3.44:1
TOP SPEED	194mph
PRODUCTION	2,378 GT3 through 2007; 1,106 GT3 RS through 2007

2002-2004 911 CARRERA AND TURBO

Porsche commemorated Y2K and model year 2000 with a Millennium Edition Carrera 4. The company limited production to 911 units and sold them worldwide; they cost DM 185,000 in Germany and $89,000 in the United States. E-Gas electronic throttles introduced on European C4 models spread to the entire range worldwide, as did PSM. Slight improvements in exhaust flow increased engine output from 296 to 300 DIN horsepower. Weissach engineers added steering wheel rocker switches for Tiptronic gear shift.

Early in calendar year 2000, Europeans took delivery of the first 996 Turbos. Using the same 3.6-liter Hans Mezger engine, Ruf and his colleagues fitted twin turbochargers and intercoolers and an improved version of Porsche's VarioCam variable camshaft timing unit, named VarioCam Plus. Turbos provided a 1.85-bar boost (27.4 psi) as low as 2,500 rpm. As engine speed increased, boost settled to 1.65 bar (24.4 psi) at 6,000 rpm, where the engine developed 420 DIN horsepower. Torque peaked at 2,700 rpm with 413 lb-ft and held that output to 4,600 rpm.

The Turbo coupe employed all-wheel drive as 993 versions had done. To accommodate new 295/30R18 rear tires, Porsche designers and engineers widened the rear of the car by 65mm (2.6 inches). As with C4 normally aspirated models, Turbo buyers had the option of the Tiptronic S transmission. PSM was standard, as was the front-end styling treatment, which was distinguished most clearly by new Bi-Xenon headlight modules. An automatic rear wing rose from the redesigned rear valence.

American customers got Turbos for model year 2001 in the summer of 2000. Oddly, a DOT misunderstanding held up the first shipments while questions of front bumper height worked their way to resolution. Porsche had set Turbo ride height 10mm lower than the C4 to improve handling. The solution that satisfied DOT was for PCNA to add small triangular black "bumperettes" slightly above the front bumper on the earliest 2001 U.S. cars. Turbos sold for DM 234,900 and $111,000 in the United States. Porsche also introduced a race-proven option for the Turbos: cross-drilled and ventilated Porsche Ceramic Composite Brakes (PCCBs), carbon fiber brake rotors impregnated with silicon carbide. Fitted with bright yellow four-piston calipers, each brake saved 5 kilograms (11 pounds) of weight from the car at its outer corners.

▲ The six-speed manual transmission was standard and provided the C4S cabriolet with acceleration from 0 to 100 kilometers per hour in 5.3 seconds. The top speed was listed as 280 kilometers (174 miles) per hour.

▶ The company introduced the Turbo-body Carrera 4S for model year 2001 in Europe. It followed with the C4S Cabriolet in Europe as a 2003 and in the United States for 2004.

▲ (Top) Porsche sold the C4S Cabriolet for €99,792 and $93,200 in the United States. The C4S coupe, with slightly quicker acceleration, went for €89,816 and $83,400.

▲ (Bottom) The M96/03 version of the 3,596cc (219.4-cubic-inch) water-cooled flat six developed 320 horsepower in the 1,565-kilogram (3,443-pound) cabriolet. A new collapsible roof system introduced for 2003 that allowed drivers to raise or lower the top at speeds up to 50 kilometers (30 miles) per hour.

YEAR	2003-2005
DESIGNATION	911 Carrera
SPECIFICATIONS	
MODEL AVAILABILITY	Coupe, Cabriolet
WHEELBASE	2350mm/92.5 inches
LENGTH	4430mm/174.4 inches
WIDTH	1770mm/69.7 inches
HEIGHT	1305mm/51.3 inches
WEIGHT	1345kg/2959 pounds (coupe) 1425kg/3135 pounds (cabriolet)
BASE PRICE	$78,146 (coupe) $88,609 (cabriolet)
TRACK FRONT	1465mm/57.7 inches
TRACK REAR	1500mm/59.1 inches
WHEELS FRONT	7.0Jx17
WHEELS REAR	9.0Jx17
TIRES FRONT	205/50ZR17
TIRES REAR	255/40ZR17
CONSTRUCTION	Unitized welded steel
SUSPENSION FRONT	Independent, light-alloy wishbones, MacPherson struts w/coil springs, gas-filled double-tube shock absorbers, anti roll bar
SUSPENSION REAR	Independent, multi-wishbone, progressive coil springs, gas-filled single-tube shock absorbers, anti roll bar
BRAKES	Ventilated, drilled discs, 4-piston aluminum monobloc calipers
ENGINE TYPE	Horizontally opposed water-cooled DOHC six-cylinder Typ M96/03
ENGINE DISPLACEMENT	3596cc/219.4CID
BORE AND STROKE	96x82.8mm/3.78x3.26 inches
HORSEPOWER	320@6800rpm (345@6800 – 40th Anniversary)
TORQUE	273lb-ft@4800rpm (all)
COMPRESSION	11.3:1
FUEL DELIVERY	Bosch DME with sequential injection
FINAL DRIVE AXLE RATIO	3.44:1
TOP SPEED	177mph
PRODUCTION	6,621 coupes; 7,254 cabriolets (both C2 and C4 total from 2002 through 2005)

▲ Because the car existed as a racing class homologation, Porsche offered the GT2 only as a rear drive model. The ceramic composite brake system (PCCB) was standard. The GT2 accelerated to 100 kilometers in 4.0 seconds and reached a top speed of 319 kilometers (198 miles) per hour.

2003-2005 911 GT2

For European customers, Porsche's rear-wheel-drive GT2 returned to the lineup using the Turbo body (though various weight reductions—primarily removing the front drive mechanisms—brought it in 100 kilograms [220 pounds] lighter than the Turbo). The Mezger 3.6-liter flat six developed 463 DIN horsepower at 5,700 rpm through twin turbos and intercoolers. As with the Turbo, the GT2 used VarioCam Plus to adjust intake valve operation and valve camshaft timing. PCCBs were standard equipment. The GT2 sold for DM 339,000 in Europe. As with the GT3, still in the lineup, Porsche offered a Clubsport option for the GT2 that delivered a competition seat, six-point harness, roll cage, and fire extinguishing system.

Alois Ruf wasted no time introducing his own R Turbo in several variations, coupe or cabriolet body in narrow or wide configurations. By altering turbochargers, fitting his own exhaust system and air intake runners, and modifying the engine management programs, his engine developed 520 DIN horsepower at 6,000 rpm. The front end incorporated vent slots and an air dam, and Ruf modified the electrically operated and adjustable rear wing. Acceleration from 0 to 62 miles per hour took 3.7 seconds, and the cars reached top speeds of 330 kilometers (205 miles) per hour or more, depending on gearing. Soon after introducing the R Turbo, he uprated its performance to 550 DIN horsepower at 6,000 rpm. This increased top speed to 351 kilometers (218 miles) per hour. A fully optioned narrow body, rear-wheel-drive coupe with integral roll cage sold for EUR 250,444 (roughly $226,000 at the time). (In hopes of developing a universal currency for its member countries, the European Union, or EU, adopted the Euro in December 1995. Banknotes and coins entered circulation on January 1, 2002.)

For 2002 the Turbo's face-lifted front styling spread across the entire 996 lineup. Standard Carrera models got a larger engine after Weissach engineers lengthened the stroke from 78mm to 82.8, a process that required considerable reengineering. In fact, according to Jürgen Barth, Weissach changed something like 80 percent of the engine components. With bore still at 96mm, overall displacement

▲ For the final year of the 996 series, Porsche upgraded the sprinter in the suit with new 18-inch alloy wheels that helped the car shed nearly 14 kilograms (31 pounds) from the previous year's model.

▼ Sport seats were optional at no extra cost, and the company deleted rear seats and the back portion of the center console to save further weight. European models developed 483 horsepower at 5,700 rpm while America customers suffered with only 477.

increased to 3,596cc (219.3 cubic inches); using VarioCam Plus, this new engine developed 320 DIN horsepower at 6,800 rpm. This was a good year for 911 enthusiasts, with seven models available. The GT2 reached U.S. shores at $179,900, as did the X50 option for the Turbos, which boosted engine output to 450 DIN horsepower at 5,700 rpm in the all-wheel-drive platform. The Carrera 2 coupe and cabriolet continued, as did the Carrera 4 cabriolet. Both cabriolet models exchanged the plastic rear window for heatable glass. A bolder, more brash C4 coupe, designated C4S, adopted the Turbo bodywork. It appeared in U.S. dealers in February 2002. Along with Turbo suspension and brakes, the C4S was 60mm (2.4 inches) wider at the rear.

YEAR	2003–2005
DESIGNATION	911 GT2
SPECIFICATIONS	
MODEL AVAILABILITY	Coupe
WHEELBASE	2355mm/92.7 inches
LENGTH	4450mm/175.2 inches
WIDTH	1830mm/72.0 inches
HEIGHT	1275mm/50.2 inches
WEIGHT	1420kg/3124 pounds
BASE PRICE	$193,700
TRACK FRONT	1495mm/58.9 inches
TRACK REAR	1520mm/59.8 inches
WHEELS FRONT	8.5Jx18
WHEELS REAR	12.0Jx18
TIRES FRONT	235/40ZR18
TIRES REAR	315/30ZR18
CONSTRUCTION	Unitized welded steel
SUSPENSION FRONT	Independent, light-alloy wishbones, MacPherson struts w/coil springs, gas-filled double-tube shock absorbers, anti roll bar
SUSPENSION REAR	Independent, multi-wishbone, progressive coil springs, gas-filled single-tube shock absorbers, anti roll bar
BRAKES	Ventilated, drilled discs, 4-piston aluminum monobloc calipers
ENGINE TYPE	Horizontally opposed water-cooled DOHC six-cylinder Typ M96/70 SL
ENGINE DISPLACEMENT	3600cc/219.7CID
BORE AND STROKE	100x76.4mm/3.94x3.01 inches
HORSEPOWER	483@5700rpm
TORQUE	472lb-ft@3500rpm
COMPRESSION	9.4:1
FUEL DELIVERY	Bosch DME with sequential injection, turbochargers, intercooler
FINAL DRIVE AXLE RATIO	3.44:1
TOP SPEED	198mph
PRODUCTION	Not available

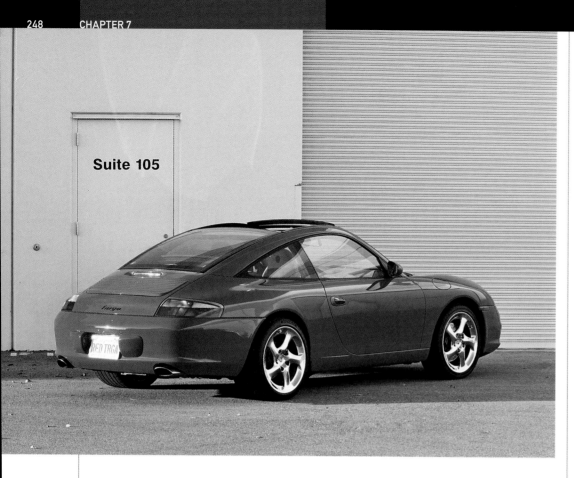

▼ As with the 993 versions, the new 996 Targa roof consisted of three elements, all of them three-layer-thick laminated safety glass tinted green. A wind deflector rose from the fixed front glass piece when the larger panel was retracted.

▲ Porsche had introduced its new 3,596cc (219.4-cubic-inch) Typ M96/03 engine for model year 2002, producing 320 horsepower at 6,800 rpm. Europeans had the option of a further upgrade to 345 horsepower through *Exclusiv*'s "Carrera Power Kit."

2001-2005 911 TARGA

At the same time, Porsche reintroduced the glass-roof Targa, which had last appeared on 993 models for 1998. The first-generation roofs occasionally stuck open, closed, or in between due to some chassis flexibility. Weissach engineers redesigned the system and its installation. The 993 system welded the top mechanism onto reinforced cabriolet bodies. The 996 assemblers brought the roof into the car through the windshield and raised it into position from inside the cabrio body before welding it in place. This method greatly increased torsional rigidity and body stiffness. Part of the new system incorporated a hatchbacklike hinged rear window, finally achieving a goal that Ferry Porsche had had for the 901 in 1964. Porsche sold the Targa for EUR 82,128 and $75,200 in the United States.

For competitors in European Carrera Cup events, Porsche Motorsports performed numerous upgrades and engine management system revisions for 2002 to boost GT3 Cup engine output to 381 DIN horsepower from the 3.6-liter Mezger engine. According to Jürgen Barth, the restyled bodywork for both road-going and racing models reduced front lift by 25 percent and rear lift by 40 percent. Motorsports developed a new seven-position rear wing. It also increased front brake rotor diameter to 350mm. For 2002 Porsche assembled 138 Cup cars, now called the Michelin Supercup following a change in series sponsorship. For 2003 engineers again improved output to 390 DIN horsepower at 7,300 rpm through a rigorous weight loss program inside the engine. Barth reported that engineers lightened pistons by 9 percent and valves by 18 percent. The 2003 GT3 Cup car weighed 1,160 kilograms (2,557 pounds). Motorsports assembled 200 of them.

For 2003 Porsche was forced to recalibrate its horsepower ratings to correspond with Society of Automotive Engineers (SAE) standards for vehicles in the United States. This resulted in a slight downward rerating, and Carrera and Carrera 4 models dropped from 320 to 315 SAE horsepower. The Turbo adjusted to 415, while the X50 package slipped from 450 to 444. The GT2 relisted at 456 SAE horsepower.

A new cabriolet top system allowed drivers to raise or lower the roof electrically while driving at speeds up to 50 kilometers (30 miles) per hour. Midway through the model year, in the spring of 2003, Porsche supplemented the Carrera 4 cabriolet with the Turbo-body Carrera 4S cabriolet, similar to the C4S coupe with the Turbo suspension, brakes, and wheels. The entire 996 lineup introduced onboard computers, a CD radio, and a new three-spoke steering wheel.

▲ (Top) The Targa weighed 1,415 kilograms (3,113 pounds). It sold for €82,276 in Germany and $76,000 in the United States.

▲ (Bottom) Porsche reintroduced the 996 Targa as a 2002 model in Europe and for 2003 in the United States. Unlike the all-glass-roof version for the 993, this new car also provided buyers an opening rear hatch that was accessible when the roof was closed.

YEAR	2001-2005
DESIGNATION	**911 Targa**
SPECIFICATIONS	
MODEL AVAILABILITY	Targa
WHEELBASE	2350mm/92.5 inches
LENGTH	4430mm/174.4 inches
WIDTH	1770mm/69.7 inches
HEIGHT	1305mm/51.3 inches
WEIGHT	1415kg/3113 pounds
BASE PRICE	$86,297 Targa
TRACK FRONT	1465mm/57.7 inches
TRACK REAR	1500mm/59.1 inches
WHEELS FRONT	8.0Jx18
WHEELS REAR	10.0Jx18
TIRES FRONT	225/40ZR18
TIRES REAR	285/30ZR18
CONSTRUCTION	Unitized welded steel
SUSPENSION FRONT	Independent, light-alloy wishbones, MacPherson struts w/coil springs, gas-filled double-tube shock absorbers, anti roll bar
SUSPENSION REAR	Independent, multi-wishbone, progressive coil springs, gas-filled single-tube shock absorbers, anti roll bar
BRAKES	Ventilated, drilled discs, 4-piston aluminum monobloc calipers
ENGINE TYPE	Horizontally opposed water-cooled DOHC six-cylinder Typ M96/03
ENGINE DISPLACEMENT	3596cc/219.4CID
BORE AND STROKE	96x82.8mm/3.78x3.26 inches
HORSEPOWER	320@6800rpm (345@6800 – 40 th Anniversary)
TORQUE	273lb-ft@4800rpm (all)
COMPRESSION	11.3:1
FUEL DELIVERY	Bosch DME with sequential injection
FINAL DRIVE AXLE RATIO	3.44:1
TOP SPEED	177mph
PRODUCTION	2.693 Targas

▲ On the center console below the radio of each Anniversary coupe, the company mounted a numbered plaque finished in GT Silver metallic and aluminum. *Photograph © 2011 David Newhardt*

▼ Porsche commemorated the 40th anniversary of the 911 introduction by offering an edition of 1,963 Carrera coupes painted in GT Silver Metallic. All cars worldwide used the "Power Kit"–equipped 345 horsepower M96/03S flat six. *Photograph © 2011 David Newhardt*

2003-2004 911 CARRERA 40TH ANNIVERSARY EDITION

Late in 2003, the company celebrated 40 years of 911 production with a 40th Anniversary Commemorative edition. Porsche limited the rear-wheel-drive model run to 1,963 examples. Offered only in GT Silver Metallic, the car utilized the latest Weissach engineering tricks to bring output to 345 SAE horsepower at 6,800 rpm. In addition to many small and large interior appointments, Porsche delivered the car with fitted luggage and a brief case in hand-tooled dark gray leather. It sold for EUR 95,616 in Europe and $89,800 in the United States.

Model year 2004 brought the long-awaited GT3 to U.S. shores, though not every variation made it to America. The car had 380 SAE horsepower, so few buyers complained. A less common edition of only 200, called the GT3 RS, was conceived as a street-legal homologation model. Based on the stripped Clubsport, these cars were sold only in Carrera white, with the buyer's choice or red or blue GT3 RS lettering on the side. Using a carbon fiber rear wing, rearview mirror housings, and front deck lid, as well as thinner, lighter rear windows, Porsche Motorsports pared an additional 50 kilograms (110 pounds) off the already lightened GT3. It was an uncompromised racing version with a loud interior, rough ride, brilliant handling, and neck-snapping acceleration and braking. Customers who bought one for street use or as collectors/speculators found it had no radio or climate control. Porsche Motorsports assembled just 68 of the RS models, as well as 48 RSR models strictly for competition. In addition, Weissach produced another 10 of the 2004 GT3 Rally models for the FIA Road Challenge series.

Competitors in the Carrera Cup saw mostly safety innovations in their cars. Revisions to the roll cage added extra X braces behind the seat that now accommodated the head and neck support (HANS) system required by the FIA. New warning lights and engine management changes alerted drivers when the engine was at upshift point, coolant was low, or a road-speed limiter held drivers to pit lane maximum speeds. Porsche sold these pure racers for EUR 109,500 (roughly $135,500), and it assembled

▲ Porsche gave the Anniversary coupes a dark gray leather interior with heated sport seats. The cars came equipped with sport suspension and Porsche Stability Management. *Photograph © 2011 David Newhardt*

YEAR	**2003-2004**
DESIGNATION	**911 Carrera 40th Anniversary**
SPECIFICATIONS	
MODEL AVAILABILITY	Coupe
WHEELBASE	2350mm/92.5 inches
LENGTH	4430mm/174.4 inches
WIDTH	1770mm/69.7 inches
HEIGHT	1305mm/51.3 inches
WEIGHT	1370kg/3014 pounds
BASE PRICE	$100,290
TRACK FRONT	1455mm/57.3 inches
TRACK REAR	1480mm/58.3 inches
WHEELS FRONT	8.0Jx18
WHEELS REAR	10.0Jx18
TIRES FRONT	225/40ZR18
TIRES REAR	285/30ZR18
CONSTRUCTION	Unitized welded steel
SUSPENSION FRONT	Independent, light-alloy wishbones, MacPherson struts w/coil springs, gas-filled double-tube shock absorbers, anti roll bar
SUSPENSION REAR	Independent, multi-wishbone, progressive coil springs, gas-filled single-tube shock absorbers, anti roll bar
BRAKES	Ventilated, drilled discs, 4-piston aluminum monobloc calipers
ENGINE TYPE	Horizontally opposed water-cooled DOHC six-cylinder Typ M96/03 S
ENGINE DISPLACEMENT	3596cc/219.4CID
BORE AND STROKE	96x82.8mm/3.78x3.26 inches
HORSEPOWER	345@6800
TORQUE	273lb-ft@4800rpm
COMPRESSION	11.3:1
FUEL DELIVERY	Bosch DME with sequential injection
FINAL DRIVE AXLE RATIO	3.44:1
TOP SPEED	180mph
PRODUCTION	1,963

about 150 of them.

Porsche's new Turbo cabriolet was far more civilized yet plenty powerful. Sharing the all-wheel-drive platform and 444-horsepower (SAE) engine with the coupe, the open car ran from 0 to 62 miles per hour in 4.3 seconds (4.9 seconds with the Tiptronic S). With its top down, it was good for a top speed of 290 kilometers (180 miles) per hour.

To characterize this series as a success is an understatement. Notwithstanding a small series of serious engine failures, primarily among early 3.4-liter engines, the first-generation water-cooled 911s were everything Porsche needed to survive and prosper into the twenty-first century. The gentrified water-cooled 996 introduced many longtime Mercedes-Benz and BMW owners to the 911 and made them converts. Some longtime loyalists criticized the car for straying from the purity of purpose that defined the 911 of the 1960s and 1970s—but without the recognition that, at the price point where Porsche's cars reside, customers expect not only performance but also electric windows, heated seats, sliding or retracting roofs that work, sophisticated audio systems they can enjoy above engine and road sounds, and climate control that coddles them no matter what goes on outside the car. The 993 came close. The 996 hit it precisely.

Because Porsche never let any good car go unimproved, the replacement, coming in 2005, addressed the comments of loyalists and new customers alike. And then Porsche turned to the creative talent at Weissach and let them loose.

▲ 2004 Ruf RGT RS Coupe
Ruf installed a full roll cage inside the car. Through liberal use of carbon fiber body panels, he brought his car's weight to 1,359 kilograms, 2,990 pounds. In homage to the original 1973 Carrera RS, Ruf added a fixed ducktail spoiler. *Photograph © 2011 David Newhardt*

▶ Acceleration from 0 to 100 kilometers took 4.1 seconds. Fender flares front and rear slightly increased wheel track, improving handling. Rear tires were 315/30R18s. *Photograph © 2011 David Newhardt*

▲ By revising the intake, creating a new exhaust system, and reconfiguring engine management programs Alois Ruf's engineers developed 395 horsepower from the normally aspired 3,600cc (219.6-cubic-inch) GT3 engine. *Photograph © 2011 David Newhardt*

2005–2008 CARRERA, CARRERA S
2006–2008 GT3, GT3 RS
2007–2008 TARGA 4, TARGA 4S
2008–2009 GT2
2008–2009 TURBO, TURBO CABRIO
2008–2010 CARRERA, CARRERA S, SPORT CLASSIC PDK, DFI
2009–2010 TURBO
2010 GT3
2009–2012/991 CARRERA, CARRERA S; CABRIO, CABRIO S
2009–2012/991 CARRERA 4, CARRERA 4S; CABRIO 4, CABRIO 4S
2009–2012/991 TARGA 4, TARGA 4S
2009–2012/991 TURBO, TURBO CABRIO
2009–2010 GT2
2010–2011 GT3, GT3 RS, CUP
2011/997–2012 CARRERA GTS, GTS CABRIO
2011 GT3, GT3RS
2012 GT3 RS 4.0
2012 CARRERA BLACK COUPE, CABRIO
2011–2012 TURBO COUPE; CABRIOLET; S COUPE, S CABRIOLET
2011/997–2012 SPEEDSTER
2011/997–2012 GT2 RS
2012 GT3 RS 4.0

CHAPTER 8

THE SIXTH GENERATION 2005–2012

2005-2008 CARRERA

Porsche switched up the development order with the 997. Weissach started with the convertible, knowing the lineup would include it, and then went to work on the coupe. "This strategy came from the engineers' point of view," August Achleitner explained. "The convertible is more difficult because of the stiffness that is necessary." Achleitner was Porsche's director of product line management for the Carrera. "Your work is easier when you consider the reinforcements from the beginning. The lead model of the Nine-Nine-Seven series was the convertible." In the past, Porsche had developed coupes first and when engineers had completed that, they started on open cars. With the 997, Achleitner's team developed both simultaneously. The 997 largely was his creation; before that time, from 1989 through 2000, he had directed new vehicle concepts and packaging for all Porsche vehicles.

Starting with the cabrios provided Weissach's engineers some unexpected benefits as they worked through target conflicts. These were the good news/bad news dilemmas arising when one decision revealed two or three more choices. Wolfgang Dürheimer, Porsche's vice president for research and development, explained the advantages the 997 derived from Achleitner's simultaneous effort.

The 911 heritage offered as many challenges to engineers and designers as it provided guidelines. Its characteristic front fenders retained a form that, as Ferry Porsche first dictated to Erwin Komenda for the 356, allowed the driver

◀ Loyalists and longtime customers welcomed back Porsche's traditional round-oval headlamps. Fenders rose more prominently than they had with the 996.

to see where the front wheels were located. The 911 carried on the iconic angled-down roofline that F. A. Porsche and his clay modelers had created. It still defined itself with the rear engine that dictated the car's shape, its handling, its sound, and its appeal.

Engineers boasted that the new 997 was 80 percent changed from the 996. More than half of that was beneath the surface. Not a single suspension piece was interchangeable. This situation flew in the face of Wendelin Wiedeking's often repeated goal of commonality of parts. But many of the new pieces were simpler to manufacture and easier to install. The 20 percent that remained unchanged included expensive elements such as the roof panel, the interior rear seats, and the 3.6-liter engine block, crankshaft, and pistons.

Nineteen-inch wheels forced powertrain engineers to rework gearboxes. Larger rolling circumferences required shorter final drive gears to take best advantage of engine torque and horsepower. Weissach engineers reexamined the manual and Tiptronic transmissions. They developed a new six-speed manual gearbox with torque and horsepower capacity to spare. Engineers increased gearshift pressure and stall speed on the five-speed Tiptronic. This gave drivers faster starts from a standstill, more powerful and spontaneous acceleration, and quicker shifts, especially in lower gears.

Engineers gave the 997 an automatic stability system known as Porsche Active Suspension Management (PASM).

Other than the 959, which many in the company considered to be a large run of prototypes, this 997 was the first time Porsche offered an electronic spring and damper system. In the cabrio, this was an industry first. Engineers substituted the coupe's front springs with coils 10 percent softer for the cabrio and substituted the coupe's rear suspension bushings with some much harder for the open car. They compressed PASM's range of variability to fit the cabriolet's slightly diminished stiffness and the anticipated character of its drivers. Its stiffest "sport" settings were roughly 15 percent softer than calibrations for the coupe, while the softest point was slightly gentler than the coupe's.

For 997 coupes, Porsche offered a full "Sport Suspension" for European customers. This lowered ride height by 20mm (0.79 inches), setting it too low to meet U.S. federal ride height standards. Engineers provided a mechanical differential lock on the rear axle of 22 percent under acceleration and 27 percent under deceleration or braking to enhance directional stability. Buyers could order it through Porsche Exclusiv as an option on the Carrera or in place of the PASM on the Carrera S. The ride was much harsher because Porsche conceived it "for the ambitious driver not so much interested in comfort but rather in super performance and agility," Achleitner explained.

Power train manager Stefan Knirsch and his staff devised intake and exhaust modifications to the familiar 3,598cc 996 engine that added 5 horsepower to reach 325 SAE horsepower for the base Carrera. This engine carried over the VarioCam Plus valve management system comprising two interacting switching cup tappets on the intake side of the engine, driven by two cams of varying size on the intake camshaft.

Enlarging cylinder bore from 96mm to 99 but retaining stroke at 82.8mm brought a new S engine displacement to 3,823cc or 3.8 liters (233.2 cubic inches). The car developed 355 SAE horsepower. Coupes reached 60 miles per hour in 4.8

▲ The company sold coupes alone for the first few months. The base Carrera 3,596cc (219.4-cubic-inch) M96/05 engine developed 325 horsepower at 6,800 rpm. The S M97/01 engine displaced 3,824cc (233.3 cubic inches) and produced 355 horsepower at 6,600 rpm.

▲ Porsche introduced the 997 in mid-2004 as a 2005 model. It offered two versions, the Carrera, with an update of the 3.6-liter engine from the 996, and the Carrera S, with a new 3.8-liter flat six.

seconds (5.2 for the cabriolet) and in 4.6 seconds for the S (but 4.9 seconds for the S cabriolet). Porsche quoted top speeds of 177 miles (283 kilometers) per hour and 182 miles (291 kilometers) per hour, respectively. This was the first time since 1977, with 2.7-liter and 3.0-liter engines, that Porsche offered two normally aspirated engines simultaneously.

Porsche introduced the 997 Carrera and Carrera S coupes in summer 2004 as a 2005 model in Europe and the United States. Through the model year, it continued manufacturing and selling 996 Turbo S and Turbo S cabriolets (installing the X50 option as standard equipment, providing the cars with 444 SAE horsepower). Cabriolet versions of the 997 Carrera and Carrera S reached dealers worldwide in April 2005. The first all-wheel-drive models arrived in showrooms in midsummer, with Targases, Turbos, GT3s, GT2s, and other models following into 2006, 2007, and 2008.

The 997 interior was nearly all new, except for the rear seats and a few details. It was the work of interior designer Anke Wilhelm and interior chief Franz-Josef Siegert. In the late 1990s, Harm Lagaay had hired Siegert away from Mercedes-Benz, and Siegert brought in a staff of designers only to do interiors and still others to attend to the details, the jewelers, as he called them. Wilhelm and Siegert's instrument pod seemed familiar to Porsche owners, and Uli Sauter's graphics were quickly comprehensible to those driving a Porsche for the first time. The 996 interior was the most comfortable and user friendly of any previous model. Porsche would not shy away from its new enthusiast base. The 997 offered four seat options to satisfy most backs and body shapes. All were comfortable and supportive, with headrests 50mm (2 inches) higher and angled closer for better support. There was more metal and less plastic in this new interior than ever before.

Forty percent of all 997s for 2005 were cabriolets, and about 50 percent of those went to the United States. The solid fixed-glass back window

▲ The new steering wheel controlled functions ranging from the navigation system, and radio tuning, to telephone functions. New seats offered optional electric side-bolster adjustment.

with an electric defogger was one of many selling points, introduced at the 996 face-lift in 2002 and slightly enlarged with the 997. At speeds up to 30 miles (50 kilometers) per hour, the 997 driver could raise or lower the roof. That process, including dropping and raising side windows, took 20 seconds in either direction.

Porsche's wholly owned subsidiary Car Top Systems (CTS) developed the mechanism for the 997's top and manufactured the complete system, with bows, hinges, motors, inner lining, glass, and outer material. The system arrived from CTS fully assembled and, like everything else on the Zuffenhausen assembly line, just in time for two men to lift it and set it onto a painted car body. The entire assembly weighed 93 pounds (42 kilograms). But weight is always an enemy to Porsche engineers. While the cabriolet gained a total of 297 pounds (135 kilograms) over the coupe, diligent management of every system kept the net weight increase to 187 pounds (85 kilograms).

Because the roofline could not exactly mimic the coupe's, engineers tweaked rear spoiler performance. It rose 20 millimeters (0.80 inches) higher on the cabriolet than on the coupe to provide more aerodynamic effect. The subtly higher wing was not the only effect that top-down Porsche drivers noticed. "The exhaust sound was even more aggressive with the cabrio," Bernd Kahnau explained with a broad grin. "Because of the open cabin, we wanted our customers to really be able to hear the engine." It reminded some drivers of the 993 more than a water-cooled 996. Because of America's relaxed exhaust noise standards, U.S. buyers got the loudest exhausts of any 997 purchasers. Bernd Kahnau knew the sound and explained the reason: "Our exhaust engineers? They are our Mozarts."

For 60 years, Porsche had listened to its engineers and designers as they continued to deliver Ferry Porsche's dream car to an ever-changing world. Dr. Erhard Mössle, the four-wheel-drive manager for the 997 Turbo, gave an example.

His team had identified "characteristics of the Targa buyer," he explained. "They desire exclusivity, distinctive design. They primarily are cruisers, you know, connoisseurs of automobiles rather than racers. And traditionally, they are safety conscious." So Porsche "repositioned" the Targa to differentiate it from those that came before and from the cabriolet and coupe.

For Mössle, whose title is product manager four wheel drive, and for his team, the decision to start with the C4 and C4S platforms immediately addressed and resolved questions about what the new Targa would be. The 44mm (1.7 inch) wider rear end provided the aggressive and dynamic form that engineers and stylist Mathias Kulla had in mind. This single decision gave them the distinctive appearance, as well as safety and stability controls, they believed appealed to potential Targa buyers. Porsche's engineers and marketers understood that these owners routinely used their cars year-round in climates with snow, so the C4 drive system made further sense. Its 10mm (0.4 inch) lower ride height didn't hurt its overall visual appeal either. But each of these factors combined to preclude any possibility of a Targa 2 or Targa 2S.

"The glass roof," Tomas Christiansen, manager of body engineering, explained, "is lighter than the glass in the 996. This makes a very big difference so high above the center of gravity. It is two kilograms [4.4 lbs] less, even with the hatchback mechanism."

The C4S, in its own right and as the basis for the Targa, was so thoroughly tamed beyond even its 996 predecessor that heavy front-end sensations that engineers anticipated never appeared. Porsche derived and developed the C4 964, introduced in 1989, with the U.S. market in mind. It tended to understeer when drivers pushed it too hard. This was a comfortable response for most Americans, and Porsche engineers—dealing also with a heavy load of equipment up front—settled for this reaction. In the 993 they compensated; some say they overcompensated. This car produced handling characteristics that were more rear dominant, reminiscent of earlier two-wheel-drive 911s. With the 996 and 997 C4 and C4S, turn-in was crisp, precise, predictable, and dependable.

Porsche began shipping its all-wheel-drive 997 C4 and C4S models to the United States in November 2005. The Turbo followed, shown first at Geneva in February 2006 along with the GT3, the other model guaranteed to quicken the purist's pulse. The 3.6-liter Mezger engine developed 480 SAE horsepower in the Turbo. Acceleration to 62 miles per hour took 3.9 seconds, and Porsche quoted its top speed as 193 miles (311 kilometers) per hour. In a testimony to Weissach's engineering, the Tiptronic version of the all-wheel-drive Turbo was even quicker, reaching 62 miles per hour in just 3.7 seconds. The Turbo cabriolet arrived in September 2007 boasting nearly identical performance figures, a longtime goal for Weissach's engineers. Marketing's tendencies for greater luxury in its flagship Turbo models continued with the latest generation. Fully optioned Turbos elevated high performance to performance art, especially if buyers optioned "paint to match" exterior color schemes and interior leather and other appointments from Porsche Exclusiv.

▲ Porsche introduced the 997 Cabriolet models in early April 2005. As with the coupes, the new open cars were available as Carrera and more potent Carrera S models.

▲ The base Carrera developed 325 horsepower. The cabriolet weighed 1,480 kilograms, 3,256 pounds. With the 997 model, Porsche made the removable hardtop an extra-cost option.

▶ (Opposite top) Engineers and designers worked to improve the operation and appearance of the collapsible fabric top when it was up. Their goal was to come closer to the shape of the coupe.

▶ (Opposite bottom) The Sport Chrono Plus feature, identified by the analog/digital stopwatch installed in the dashboard, added significant performance potential to the Carrera S models. Sports Chrono allowed drivers to switch engine management programs to improve driving enjoyment over repeated roads.

▲ To honor Porsche's long relationship with PCA, the Porsche Club of America, Weissach designers, engineers, and planners created a "Club" coupe painted in the PCA logo blue for the 2006 model year.

▲ (Top) The Carrera S rode on 235/35ZR19 front tires and 295/30ZR19s on the rear. The coupe weighed 1,420 kilograms (3,124 pounds). Sport Suspension lowered ride height by 20mm (0.8 inches).

▲ (Bottom) Carrera S coupes accelerated from 0 to 100 kilometers per hour in 4.8 seconds. Porsche quoted a top speed of 293 kilometers (182 miles) per hour.

YEAR	2005-2008
DESIGNATION	911 Carrera
SPECIFICATIONS	
MODEL AVAILABILITY	Coupe, Cabriolet
WHEELBASE	2350mm/92.5 inches
LENGTH	4427mm/174.3 inches
WIDTH	1808mm/71.2 inches
HEIGHT	1310mm/51.6 inches
WEIGHT	1395kg/3069 pounds
BASE PRICE	$69,300 coupe - $79,100 cabriolet
TRACK FRONT	1486mm/58.5 inches
TRACK REAR	1534mm/60.4 inches
WHEELS FRONT	8.0Jx18
WHEELS REAR	10.0Jx18
TIRES FRONT	235/40ZR18
TIRES REAR	265/40ZR18
CONSTRUCTION	Monocoque steel
SUSPENSION FRONT	Independent, wishbones, semi-trailing arms, MacPherson struts w/coil springs, gas-filled double-tube shock absorbers, anti roll bar
SUSPENSION REAR	Independent, multi-wishbone, progressive coil springs, gas-filled single-tube shock absorbers, anti roll bar
BRAKES	Ventilated, drilled discs, 4-piston aluminum monobloc calipers
ENGINE TYPE	Horizontally opposed water-cooled DOHC six-cylinder Typ M96/05
ENGINE DISPLACEMENT	3596cc/220.0CID
BORE AND STROKE	96x82.8mm/3.78x3.26 inches
HORSEPOWER	325@6800rpm
TORQUE	273lb-ft@4250rpm
COMPRESSION	11.3:1
FUEL DELIVERY	Bosch DME with sequential injection
FINAL DRIVE AXLE RATIO	3.44:1
TOP SPEED	178mph
PRODUCTION	coupe: 6,239; cabriolet 3,019

2006-2008 GT3

The GT3 appeared alongside the Turbo in Geneva in 2006. This normally aspirated version of the Mezger 3.6-liter flat six achieved 415 SAE horsepower in this incarnation, giving owners acceleration to 60 miles per hour in 4.1 seconds and a top speed of 193 miles (311 kilometers) per hour. Project leader Andreas Preuninger explained the genesis and evolution of the car, still a homologation "special": "It's as close as you can get to a race car with a license plate on it. We wanted to translate the feeling a race car gives you, the emotion, the wish to drive to your destination in a circle, not in a straight line, because you don't want to get out of the car!"

In 2007 Porsche offered its GT2, Preuninger's rear-drive 530-horsepower (SAE) 911 that reached 204 miles (328 kilometers) per hour, the company's first production street-legal 911 to exceed 200 miles per hour. This large horsepower jump came about by using twin turbochargers and a newly designed expansion intake manifold. The "distributor" pipe was longer than that used on the Turbo, while the intake manifolds were shorter. Weighing 3,175 pounds (1,440 kilograms), the car reached 62 miles per hour in 3.6 seconds. Ironically, because the GT2 was another Porsche homologation model for a rear-drive racing category, acceleration to 62 miles per hour was faster in the all-wheel-drive Tiptronic Turbo than the more powerful rear-wheel-drive GT2.

The GT2 reached U.S. buyers as a 2008 model following its debut in November 2007 at the Los Angeles Auto Show. Porsche's release materials explained that the car was "meant for the driver who desires exclusivity coupled with race-track capable handling and acceleration." Acceleration was a key issue. "All GT2 drivers," the release continued, "should be able to match those acceleration times consistently thanks to the debut of the Porsche Launch Assist which sets engine speed and turbo boost for optimum acceleration."

Not only was the GT2 quick and fast, it also was environmentally friendly, classified as a low-emissions vehicle (LEV-II). Porsche even offered an optional Bose Surround Sound audio system with 13 speakers, a seven-channel digital amplifier, and 325 watts of total output. It was a highly civilized race car for the road.

For those seeking a highly civilized, fast, quick open car with greater creature comfort, Porsche introduced its cabriolet version of the 997 Twin Turbo. The 3.6-liter engine developed 480 SAE horsepower at 6,000 rpm. PSM and PASM systems were standard equipment. Using these, the strengthened Tiptronic S transmission, and Sport Chrono Plus with Porsche Launch Assist, the open-top Turbo accelerated to 62 miles per hour in 3.5 seconds, a tenth of a second quicker than the GT2, thanks to all-wheel-drive traction and automatic shifting. Porsche listed top speed at 193 miles (309 kilometers) per hour. The company installed a Bose

sound system and leather upholstery as standard equipment on the Turbo coupe and cabriolet.

For its open car to accommodate the horsepower, torque, and handling challenges it was capable of enduring, Weissach engineers spot welded and bonded the floor and side panels to the platform. They also welded additional side reinforcement plates to the sills, to triangular joint plates at the rear of the door pillars, and to the A-pillar to further stiffen the Turbo cabriolet.

The carryover GT3 and the new GT3RS both used the Hans Mezger–designed normally aspirated 3.6-liter dry sump engine, which developed 415 SAE horsepower at 7,600 rpm. The RS version took the GT3's racing heritage further for those drivers who spent time on racetracks as well as public roads. Its 44mm (1.73-inch) wider rear body provided room for the split rear suspension wishbones, allowing for higher lateral grip. PASM was standard equipment on the RS, as was the GT2's asymmetrical limited slip differential. The traction control system incorporated automatic brake differential, automatic slip control, and engine drag control. Both the RS and the GT3 shared other features, including Alcantara-upholstered seats and an Alcantara-trimmed steering wheel with a yellow band at the top to indicate when front wheels were aimed straight ahead.

◀ (Opposite) The company introduced the 997 versions of its GT3 in late February 2006 at the Geneva Auto Show. Weissach engineers went back to basics with the engine and eventually tuned the 3,600cc (2319.6-cubic-inch) flat six to develop 415 horsepower at 7,600 rpm. *Photograph courtesy Porsche Cars North America*

▲ (Top) Michelin Supercup racer Markus Winkelhock took television viewers for a season of great rides, seen through the T-shaped camera mount on his roof. During the 2006 season, the Supercup cars developed 400 horsepower and weighed 1,150 kilograms, 2,535 pounds. *Photograph courtesy Porsche Cars North America*

▲ (Bottom) The more aggressive-looking GT3 RS incorporated 44mm (1.73-inch)-wider rear fenders, a plastic back window, and an adjustable rear wing. It weighed 1,375 kilograms (3,031 pounds) and with its 415 horsepower engine, accelerated from 0 to 100 kilometers per hour in 4.2 seconds. *Photograph courtesy Porsche Cars North America*

▲ 2009 997 GT3 RSR Coupe

Porsche developed this car to be eligible in several racing series, each with slightly different rules. FIA regulations allowed the car to race as light as 1,200 kilograms, 2,646 pounds.

▶ Weissach racing engineers welded in 30 meters of seamless steel tubing to form the integral roll cage. The transmission was Porsche's six-speed sequential gearbox that created what felt like in-line shifting without the need to move the lever left or right.

▲ (Top) The new M97/80 engine, with bore of 102.7 millimeters (4.04 inches) and stroke of 76.4 millimeters (3.01 inches) displaced exactly 3,800cc (231.8 cubic inches) and developed 465 horsepower at 8,000 rpm. Weissach assembled 37 of these cars.

▲ (Bottom) Bolted-on fender extensions widened the car by 50mm (1.97 inches) on each side. The car rode on 11-inch wheels at the front and 13s in the rear. Tire sizes were 27/65-18 fronts and 31.71-18 rears.

YEAR	1999-2005
DESIGNATION	911 GT3/GT3RS
SPECIFICATIONS	
MODEL AVAILABILITY	Coupe
WHEELBASE	2350mm/92.5 inches
LENGTH	4430mm/174.4 inches
WIDTH	1765mm/69.5 inches
HEIGHT	1305mm/51.3 inches
WEIGHT	1350kg/2970 pounds
BASE PRICE	$345,000
TRACK FRONT	1471mm/57.3 inches
TRACK REAR	1490mm/59.1 inches
WHEELS FRONT	10.0Jx18
WHEELS REAR	11.0Jx18
TIRES FRONT	295/40-18
TIRES REAR	335/30-18
CONSTRUCTION	Unitized welded steel
SUSPENSION FRONT	Independent, light-alloy wishbones, MacPherson struts w/coil springs, adjustable gas-filled double-tube shock absorbers, anti roll bar
SUSPENSION REAR	Independent, multi-wishbone, progressive coil springs, adjustable gas-filled single-tube shock absorbers, anti roll bar
BRAKES	Ventilated, drilled discs, four-piston aluminum monobloc calipers
ENGINE TYPE	Horizontally opposed water-cooled DOHC six-cylinder Typ M96/77
ENGINE DISPLACEMENT	3598cc/206.7CID
BORE AND STROKE	100x76.4mm/3.78x3.07 inches
HORSEPOWER	420@8200rpm
TORQUE	288lb-ft@7000rpm
COMPRESSION	11.7:1
FUEL DELIVERY	Bosch DME with sequential injection
FINAL DRIVE AXLE RATIO	Varies
TOP SPEED	189mph
PRODUCTION	51

2007-2010 911 TARGA AND CARRERA

Carrera and Carrera S models and their all-wheel-drive sibling Carrera 4 and 4S versions received upgrades as well. Porsche offered customers for coupe, cabriolet, and Targa models leather sports bucket seats, basically a racing seat with a folding backrest for easier passenger or luggage access to the rear of the cabin. An X51 Power Kit upgraded S engine output from 355 SAE horsepower at 6,600 rpm to 381 horsepower. PASM continued as standard equipment on S-level models and remained optional on base Carrera, Carrera 4, and Targa 4 cars. The PCCB system was available across the lineup.

Two significant engineering developments characterized the 2009 model year 997 series—features so long awaited and offering such performance potential that enthusiasts referred to the new cars as 997/2, or 997 second-generation models. The first development was two new engines using new two-piece crankcases, revised intake and exhaust systems, an updated VarioCam Plus intake valve timing and lift system, and a new fuel injection system directly into each cylinder. The 3.6-liter engine, with bore of 97mm and stroke of 81.5, displaced 3,614cc (220.5 cubic inches), while the new 3.8-liter S power plant, with 82.8mm bore and 77.5mm

▼ At introduction, Porsche priced the Targa 4 at €79,000 in Europe and $99,700 in the United States. This Targa 4S model sold for €87,900 on the continent and $110,900 from U.S. dealers.

◀ The third-generation glass-roof Targa appeared in November 2006. Following a great deal of research, Porsche decided to assemble the car only on all-wheel-drive Carrera 4 and 4S platforms.

▼ Twin electric motors took seven seconds to open the two-layer tinted glass roof panel. When it was fully retracted, it provided 0.45 square meters (4.8 square feet) of opening.

stroke, displaced 3,800cc (231.8 cubic inches) exactly. Taken all together, these changes produced more power; the base Carrera 3.6-liter engine rose from 325 SAE horsepower to 345 at 6,500 rpm and from 273 to 288 lb-ft torque at 4,400. The S increased from 355 to 388 SAE horsepower at 6,500 rpm, and torque went from 295 lb-ft to 310 at 4,400 rpm. Engineers reduced each engine's weight by 6 kilograms (13.2 pounds) over its previous version. Fuel economy and hydrocarbon emissions improved for both engines.

Porsche also introduced its long-awaited seven-speed PDK "manumatic" transmission. The PDK carried over nothing from racing versions of the PDK from the 1980s. Those gearboxes were designed, calibrated, and assembled to endure countless cycles of full-throttle acceleration and extreme braking, techniques seldom seen in series production cars. So Porsche waited until computers and electronics were able to provide desired responses in moderate to severe acceleration, cruise control long-distance driving, coasting, and other real-world conditions. Using two clutches and what is essentially two half gearboxes, the computer preselected the next anticipated gear and activated the change up or down with no interruption in power delivery. A console-mounted lever provided the familiar forward upshift/backward downshift driving style. Sliding switches on the steering wheel operated the transmission similarly. Drivers also could ignore the system completely and run the car as one with a fully automatic shifting transmission. Michael Niko, development engineer for the PDK, made it clear that Porsche would continue to develop, improve, and market its six-speed manual gearboxes; these were needed for racing homologation.

Porsche also provided the cars more potent brakes. All four rotors were cross drilled and inner ventilated and measured 330mm (12.99 inches) in diameter. Both base Carrera and Carrera S models received the four-piston mono-block calipers used on 911 Turbo models.

▲ **2009 997 Carrera Coupe**
Direct fuel injection into new water-cooled engines with 3,614cc (220.5-cubic-inch) displacement brought output up to 345 horsepower at 6,500 rpm for the base Carrera. The long-awaited seven-speed Porsche Double-Clutch transmission (Porsche *Doppelkupplungsgetriebe*, or PDK) or the six-speed manual were available. *Photograph Courtesy Porsche Cars North America*

Externally, Porsche designers replaced taillights and front marker fixtures with light-emitting diodes (LEDs) with crisp on-off illumination. Bi-Xenon low and main headlight beams replaced the previous lamps, supplemented by an optional system that swiveled low beams in response to steering inputs. Beneath the headlights, larger air intakes provided better cooling airflow to the side-mounted radiators. These were so efficient that they eliminated the need for the low center-mounted cooler for PDK-equipped cars.

The Porsche Communication Management (PCM) system provided a 12 percent larger touch screen—165mm (6.5 inches) versus 147mm (5.8 inches). An optional navigation module incorporated a 40-gigabyte hard drive, operable by voice command. It included an internal GSM phone with Bluetooth hands-free operation and provided connectivity to external music sources from iPods or USB memory sticks.

▼ The Carrera S coupe weighed 1,455 kilograms (3,208 pounds). The PDK added 30 kilograms (66 pounds) to the car's weight. The 3,800cc (231.8-cubic-inch) S engine developed 385 horsepower at 6,500 rpm. *Photograph Courtesy Porsche Cars North America*

◀ A subtle trim strip of anodized and polished aluminum ran along the roofline above the window glass to highlight the Targa model. With the large glass panel closed, owners could open the rear hatch to access behind-the-seat luggage space.

▲ The Carrera S cabriolet weighed 1,510 kilograms (3,329 pounds) with a six-speed manual gearbox. Porsche published a top speed of 302 kilometers (188 miles) per hour with either transmission. The car rode on 235/35ZR19 front tires and 295/30ZR19s on the rear. *Photograph Courtesy Porsche Cars North America*

▲ (Top) While technologically this car was a subtle advance on the fine art of the Carrera S, visually it was a reminder of company history. The rear ducktail recalled the 1973 RS, and the car's only color, Sport Classic Grey, came from the 356. *Photograph Courtesy Porsche Cars North America*

▲ (Bottom) Porsche tuned the Sport Classic engine—with Performance Kit—to develop 408 horsepower. Developed by Porsche *Exclusiv* to show its capabilities to a wider audience, the special division assembled just 250 at €169,300 each (roughly $224,000), none of which were destined to U.S. shores.

YEAR	2007–2008
DESIGNATION	911 Targa 4, Targa 4S
SPECIFICATIONS	
MODEL AVAILABILITY	Targa
WHEELBASE	2350mm/92.5 inches
LENGTH	4427mm/174.3 inches
WIDTH	1852mm/72.9 inches
HEIGHT	1310mm/51.6 inches
WEIGHT	1450kg/3190 pounds 1475kg/3245 pounds (4S)
BASE PRICE	Not available
TRACK FRONT	1488mm/58.5 inches
TRACK REAR	1548mm/59.8 inches
WHEELS FRONT	8.0Jx18 – 8.0Jx19 (4S)
WHEELS REAR	11.0Jx18 – 11.0Jx19 (4S)
TIRES FRONT	235/40ZR18 – 235/35ZR19 (4S)
TIRES REAR	295/35ZR18 – 305/30ZR19 (4S)
CONSTRUCTION	Monocoque steel
SUSPENSION FRONT	Independent, wishbones, semi-trailing arms, MacPherson struts w/coil springs, gas-filled double-tube shock absorbers, anti roll bar
SUSPENSION REAR	Independent, multi-wishbone, progressive coil springs, gas-filled single-tube shock absorbers, anti roll bar
BRAKES	Ventilated, drilled discs, 4-piston aluminum monobloc calipers
ENGINE TYPE	Horizontally opposed water-cooled DOHC six-cylinder M96/05 – M97/01 (4S)
ENGINE DISPLACEMENT	3596cc/220.0CID – 3824cc/233.3CID (4S)
BORE AND STROKE	96x82.8mm/3.78x3.26 inches 99x82.8mm/3.90x3.26 inches (4S)
HORSEPOWER	325@6600rpm – 355@6600rpm (4S)
TORQUE	273lb-ft@4250rpm – 295lb-ft@4600rpm (4S)
COMPRESSION	11.3:1 – 11.8:1 (4S)
FUEL DELIVERY	Bosch DME with sequential injection
FINAL DRIVE AXLE RATIO	3.44:1
TOP SPEED	175mph Targa 4; 180mph Targa 4S

2008–2009 GT2

From Porsche new engines with direct fuel injection and PDK transmissions redefined two of its benchmarks for 2010. The 997 second-generation 3,800cc flat six replaced the tried-and-true 3.6-liter Mezger dry sump power plant and powered the GT3 and GT3RS models starting in 2010. It delivered 430 SAE horsepower at 7,600 rpm in the GT3 and 450 at 8,500 in the RS version. Both cars were available only with upgraded six-speed manual transmissions. The second-generation 3.8-liter engine found its way across more of the high-performance lineup, appearing in the Turbo and the GT2 models, as well as in the normally aspirated second-generation GT3 RS.

If one premise of twenty-first-century motoring was zero carbon emissions, Alois Ruf set the benchmark with an all-electric 911 known internally as Project A and in the magazine world as e-Ruf. An idea born in his Pfaffenhausen shops in late 2006, it recalled a family heritage that included building small hydroelectric plants along a couple of Bavarian rivers. Ruf's Project A was a 997 cut spare as could be, with few slats or styling cues. But battery weight is an unavoidable reality; it carried 1,200 pounds (545 kilograms) of batteries. His seven-figure prototype was and is a prototype, and like everything Alois Ruf did, it inspired (and sometimes enraged) Weissach. The small size of his firm allowed him to respond to and act on new ideas and concepts more quickly.

▲ Acceleration from 0 to 100 kilometers per hour took 3.7 seconds. Porsche charged €159,100 in Europe, about $233,500 at the time. Porsche Carbon Composite Brakes were standard equipment.

▶ (Opposite) Despite its gentrified appearance, it was a race car in slight disguise. Beneath the carpets were six mounting points for a roll cage. An assortment of six airbags—same as in any Carrera or Carrera S model—was installed as well.

▲ The GT2 rode about 25mm (1 inch) lower than base Carrera. Every piece of the suspension was tuned for stiffer ride and superior handling so the car remained neutral to its limit of cornering adhesion.

THE SIXTH GENERATION 2005–2012

▲ This was 3,600cc (219.6 cubic inches) put to efficient use with the aid of two turbochargers. Output was 530 horsepower at 6,500 rpm. With a top speed of 326 kilometers (204 miles) per hour, it was Porsche's first 911 to cross that speed threshold.

YEAR	2008-2009
DESIGNATION	911 GT2
SPECIFICATIONS	
MODEL AVAILABILITY	Coupe
WHEELBASE	2360mm/92.9 inches
LENGTH	4469mm/175.9 inches
WIDTH	1852mm/72.9 inches
HEIGHT	1285mm/50.6 inches
WEIGHT	1440kg/3168 pounds
BASE PRICE	$191,700
TRACK FRONT	1515mm/59.6 inches
TRACK REAR	1550mm/61.0 inches
WHEELS FRONT	8.5Jx19
WHEELS REAR	12.0Jx19
TIRES FRONT	235/35ZR19
TIRES REAR	325/30ZR19
CONSTRUCTION	Monocoque steel
SUSPENSION FRONT	Independent, wishbones, semi-trailing arms, MacPherson struts w/coil springs, gas-filled double-tube shock absorbers, anti roll bar
SUSPENSION REAR	Independent, multi-wishbone, progressive coil springs, gas-filled single-tube shock absorbers, anti roll bar
BRAKES	Ventilated, drilled discs, 4-piston aluminum monobloc calipers
ENGINE TYPE	Horizontally opposed water-cooled DOHC six-cylinder M97/70 S
ENGINE DISPLACEMENT	3600cc/219.6CID
BORE AND STROKE	100x76.4mm/3.94x3.01 inches
HORSEPOWER	530@7600rpm
TORQUE	502lb-ft@2200-4500rpm
COMPRESSION	9.0:1
FUEL DELIVERY	Bosch DME with sequential injection, turbochargers, intercooler
FINAL DRIVE AXLE RATIO	3.44:1
TOP SPEED	206mph

2008-2009 911 TURBO

The new Turbo (in both coupe and cabriolet forms) gave buyers astonishing performance, thanks to the new engine and to updated variable turbine geometry that increased torque and improved low-end response by varying turbine guide blade angles to reduce turbo lag at low engine speeds. The 3.8-liter engine, a derivative of the Carrera S power unit that resulted primarily from changes in casting processes, developed 500 SAE horsepower at 6,000 rpm, 20 more than its predecessor, and 479 lb-ft of torque between 1,950 and 5,000 rpm, an increase of 22 lb-ft. By using more efficient intercoolers and increasing the compression ratio from 9.0:1 to 9.8:1, Weissach engineers reduced boost from 1.0 atmosphere (14.8 psi) to 0.8 (11.84 psi). As a result, acceleration from 0 to 62 miles per hour required just 3.1 seconds. Acceleration to 100 miles per hour took 6.9 seconds, and the cars reached a top speed of 194 miles (310 kilometers) per hour, all while reducing CO_2 emissions and fuel consumption. To achieve those acceleration times, buyer had to opt for both the seven-speed PDK transmission and the Sport Chrono Plus system. To enhance handling, Weissach engineers created the Porsche Active Drivetrain Mount (PADM), an electromagnetic engine mount that grew stiffer to reduce undesirable engine movement under higher cornering loads. The manual six-speed may have delivered greater driving pleasure for some owners, but the PDK outraced them. Porsche sold the Turbo coupe for $132,800 and the cabriolet for $143,800.

▲ Variable turbine geometry almost eliminated turbo lag. The 3,600cc (219.6-cubic-inch) water-cooled engine developed 480 horsepower at 6,000 rpm.

▲ (Top) Porsche quoted acceleration from 0 to 100 kilometers as taking less time with the Tiptronic (3.7 seconds) than the six-speed manual (4.0). Top speed was the same with either gearbox at 307 kilometers (192 miles) per hour.

▲ (Bottom) As *Excellence* editor Pete Stout put it, "This won't give you the hair-raising thrills a GT3 will, but it isn't meant to…. The Turbo cabriolet's chassis feels beautifully judged in a way the coupe's just doesn't."

◀ The interior was classic, comfortable, and familiar to more than 100,000 buyers of the 997 by the end of the 2007 model year. Leather seating and the DVD-based navigation system, Porsche Communication Manager, all were standard.

▲ The instrument panel was easily recognized. The Tiptronic five-speed shifted from small rocker switches on the steering wheel or the traditional center console position.

YEAR	2008-2009
DESIGNATION	911 Turbo
SPECIFICATIONS	
MODEL AVAILABILITY	Coupe, Cabriolet
WHEELBASE	2360mm/92.9 inches
LENGTH	4450mm/175.2 inches
WIDTH	1852mm/72.9 inches
HEIGHT	1285mm/50.6 inches
WEIGHT	1585kg/3487 pounds
	1655kg/3641 pounds cabriolet
BASE PRICE	$126,200 coupe - $136,500 cabriolet
TRACK FRONT	1490mm/58.7 inches
TRACK REAR	1550mm/61.0 inches
WHEELS FRONT	8.5Jx19
WHEELS REAR	11.0Jx19
TIRES FRONT	235/35ZR19
TIRES REAR	305/30ZR19
CONSTRUCTION	Monocoque steel
SUSPENSION FRONT	Independent, wishbones, semi-trailing arms, MacPherson struts w/coil springs, gas-filled double-tube shock absorbers, anti roll bar
SUSPENSION REAR	Independent, multi-wishbone, progressive coil springs, gas-filled single-tube shock absorbers, anti roll bar
BRAKES	Ventilated, drilled discs, 4-piston aluminum monobloc calipers
ENGINE TYPE	Horizontally opposed water-cooled DOHC six-cylinder M97/70
ENGINE DISPLACEMENT	3600cc/219.6CID
BORE AND STROKE	100x76.4mm/3.94x3.01 inches
HORSEPOWER	457@6500rpm
TORQUE	502lb-ft@1950-5000rpm
COMPRESSION	9.0:1
FUEL DELIVERY	Bosch DME with sequential injection, turbochargers, intercooler
FINAL DRIVE AXLE RATIO	3.44:1
TOP SPEED	206mph

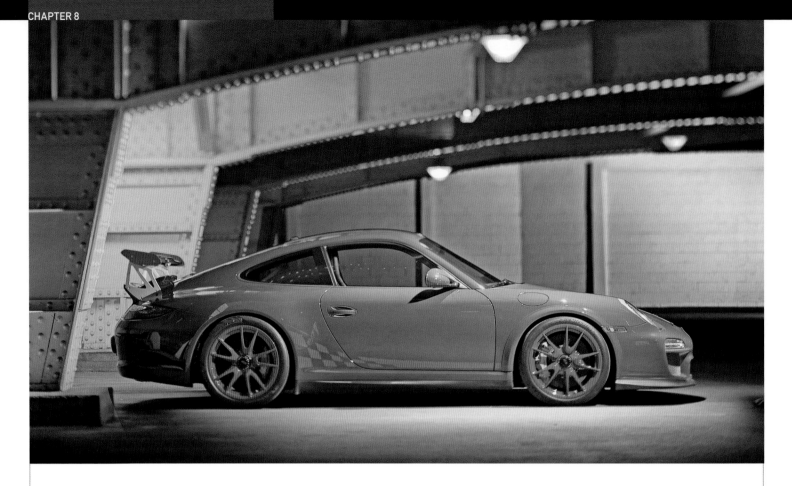

▲ The RS rode on 245/35ZR19 tires in front and 325/30ZR19s on the rear. It weighed 1,370 kilograms, 3,020 pounds. The cars were available only with the six-speed manual transmission. *Photograph © 2011 Dave Wendt*

2010 GT3

The GT3 RS was available in just three colors, Grey Black, Carrera White, and Aqua Blue Metallic, with white, gold, or red graphics and a color-keyed front grille, mirrors, wheels, and rear wing end pieces. Revising the intake manifold and raising the compression ratio gave the engine an extra 15 horsepower for a total of 450 SAE horsepower, good to get the car from standstill to 62 miles per hour in 4.0 seconds. In standard trim, the European-specification RS weighed 3,070 pounds (1,395 kilograms), but options could reduce the weight by 70 pounds (32 kilograms) or more. Europeans were offered only a Club Sport version, with lightweight fixed bucket seats, an onboard fire system, and a roll cage. American dealers got $132,800 for the GT3 RS. RSR and Cup variations followed in quick succession.

Porsche entered a unique prototype in the opening event of the 10-race series VLN (Verenstaltergemeinschalf Langstreckenpokal Nürburgring), run over the 16-mile *nordschliefe* (north-loop) circuit. The series, founded in 1977, runs one-day events on Saturdays, with practice, qualifying, and the race all on the same day. As many as 30 classes in four divisions compete in events lasting as long as 24 hours. At the 'Ring, Porsche ran eight cars under the management of Manthey Racing, including the 997 GT3 R Hybrid, which Porsche debuted at the Geneva Auto Show. Painted silver and orange, the car incorporated two 82-horsepower electric motors powered by kinetic energy absorbed under braking and stored in a flywheel housing mounted where the passenger seat would be. Running as high as 40,000 rpm, this fed energy to the front wheels, increasing acceleration with no additional fuel consumption.

One other non–North American market Carrera S came from Porsche Exclusiv. This was the 911 Sport Classic. Installed in the wide-body C4/C4S and available only in non-metallic Sport Classic Grey, it represented, according to director Ingo Frankel, an homage to several models from the company's history, including the 1973 RS, with a small fixed *bürzel* rear wing, a sculpted "double-bubble" roof profile from some of the earliest racers, and black center Fuchs-style wheels. Sport PASM and PCCB were standard. A new intake manifold increased engine output to 408 SAE horsepower. The Sport Classic came only with the six-speed manual gearbox. Porsche sold 250 copies for EUR 169,300 (about $250,000, though it was not available to North American customers).

Porsche introduced two cars of its own for those who found the Turbo a

▲ Porsche's 3,797cc (231.6-cubic-inch) flat six developed 450 horsepower in the GT3 RS. The company quoted acceleration from 0 to 100 kilometers per hour in 3.9 seconds and a top speed of 308 kilometers (193 miles) per hour. *Photograph © 2011 Dave Wendt*

YEAR	2010
DESIGNATION	911 GT3, GT3 RS
SPECIFICATIONS	
MODEL AVAILABILITY	Coupe
WHEELBASE	2355mm/92.7 inches
LENGTH	4481mm/176.4 inches
WIDTH	1808mm/71.2 inches
	1852mm/72.9 inches (RS)
HEIGHT	1280mm/50.4 inches
WEIGHT	1395kg/3069 pounds
	1370kg/3020 pounds (RS)
BASE PRICE	Not available
TRACK FRONT	1497mm/58.9 inches
TRACK REAR	1524mm/60.0 inches
WHEELS FRONT	8.5Jx19 – 9Jx19 (RS)
WHEELS REAR	12.0Jx19 – 12Jx19 (RS)
TIRES FRONT	235/35ZR19 – 245/35ZR19 (RS)
TIRES REAR	305/30ZR19 – 325/30ZR19 (RS)
CONSTRUCTION	Monocoque steel
SUSPENSION FRONT	Independent, double wishbones, trailing links, divided control arm; monotube shock absorber with progressive coil springs
SUSPENSION REAR	Independent, 5 control arms, adjustable camber, monotube shock absorber with coil springs
BRAKES	Ventilated, drilled discs, 6-pison front 4-piston rear aluminum monobloc calipers
ENGINE TYPE	Horizontally opposed water-cooled DOHC six-cylinder M97/76
ENGINE DISPLACEMENT	3797cc/231.7CID
BORE AND STROKE	103x76.4mm/4.04x3.01 inches
HORSEPOWER	435@7600rpm
TORQUE	317lb-ft@6250rpm
COMPRESSION	12.0:1
FUEL DELIVERY	Direct fuel injection
FINAL DRIVE AXLE RATIO	3.44:1
TOP SPEED	194mph

bit short on performance. First came the rear-wheel-drive GT2 RS, weighing just 1,370 kilograms and powered by the Mezger 3.6-liter flat six tuned and modified to deliver 620 SAE horsepower at 6,500 rpm. It was one of Porsche's lightest 911s ever, and its most powerful so far. It was a no apologies and no compromises road car meant more for track use than any other purpose. What began as a kind of "hobby project" for Porsche Motorsports engineer Andreas Preuninger and his colleagues to see how much further they could take a GT2 turned into a severe exercise in weight reduction to an already lightweight automobile. Plastic side and rear windows and more extensive use of carbon fiber on body panels led the list. With engine modifications to accommodate 1.6 bar (23.7 psi) of boost, output far eclipsed the GT2 figure of 530 horsepower. Porsche produced 500 of the cars.

▲ The water-cooled 3,800cc (231.8-cubic-inch) flat six developed 500 horsepower at 6,000 rpm. With all-wheel drive, the PDK transmission and optional Sports Chrono Plus, the Turbo was Porsche's fastest accelerating sports car from 0 to 100 kilometers taking only 3.3 seconds. *Photograph Courtesy Porsche Cars North America*

2009-2010 TURBO

To answer the call for more from the Turbo, the company introduced the next-generation Turbo S, which put 530 SAE horsepower under the control of all-wheel-drive technology and transmitted it to the road only through the seven-speed PDK with new steering wheel–mounted paddle shifters, the gearbox Porsche engineers believed capable of reliably channeling that kind of power output to all four wheels for the life of the car. This fourth-generation S variation of the Turbo was the best to that point, not only for its impressive performance statistics but for other details, including standard equipment ceramic composite brakes, center-lock wheels, Porsche torque vectoring (which applied braking pressure to individual rear wheels to increase the turning effect), dynamic engine mounts, Sports Chrono Plus, adaptive sport seats, a handsome carbon fiber air intake box, and dynamic bending lights that followed the front steering tires around corners. It was a package—and a packaging masterpiece—that 911 development chief August Achleitner predicted would outsell the regular Turbo by three to one.

To those most discriminating Porsche customers for whom even 250 models of a 911 Sport Classic represented too much chance of seeing a twin on the road, Porsche Exclusiv and American Express teamed up to create the 2010 Porsche 911 Centurion Edition. Inspired by F. A. Porsche's original 901 shape, the ultralimited edition celebrated American Express' 10 anniversary of the black titanium credit card. Based on the Carrera S, the Centurion Edition was tweaked to develop 385 SAE horsepower, driven through the PDK, and it was the first non-Turbo PDK to introduce paddle shifters. The car, painted in Black Metallic, was finished with a black leather, Alcantara, and carbon fiber interior. Centurion card membership required $250,000 in annual expenditures on the card, and it is safe to assume that each of the three U.K. customers who acquired one of these black 911s made a dent in his or her spending requirement.

As model year 2010 drew to a close, Porsche announced that it would add two new variations to its lineup for 2011, a 911GTS, combining the best of the GT3 and Carrera S models, and a Speedster, inspired by the 356s and carrying on styling cues of the 1989 and 1994 models. For those looking further out, rumors solidified that the next-generation 911, internally known as Typ 991, was well along in development and headed for debut as a 2012 model.

Boardroom politics and financial machinations that kept Porsche on or near the front pages of the world's financial journals during 2008, 2009, and early 2010 seemed, in the end, to have little impact on the products the company created, developed, and sold.

Conversations with Porsche management suggested that future generations of Porsche 911 cars might develop less horsepower than what enthusiasts enjoyed at the end of 2010. But no one inside or outside the company expected this to mean the cars would lose performance. Instead, Weissach engineers continued to work to the requirements of the new company motto: "Intelligent Performance." That concept put regenerating hybrid motors in GT3R race cars, electric motors in wheel hubs of the 918 prototypes, and diesel engines in Cayenne and Panamera models. That concept also shed dozens of kilograms from already lightweight GT2 and GT3 models to create RS variations and continued to do so for Carrera and Carrera S models, as well as Turbos.

Outside the company, individual entrepreneurs—from Rainer and Dieter Buchmann to Ekkehard Zimmerman and Uwe Gemballa, from longtime innovator Alois Ruf to recent entry Jon Fatthauer with his 9ff Porsches from Dortmund—have pushed and pulled Porsche and its 911 since the 1970s. By virtue of their size and their client bases, they can respond to new technology and ideas more quickly. Tuners come and go, some suffering ignominious ends, such as Gemballa's in 2010. Others, like the Buchmanns' b+b and Zimmerman's dp, are less dramatic. Creative inventors such as Ruf survive for decades because, to paraphrase Ferry Porsche's motive for creating his 356, they look around and see that no one has done what they want. No one studying Porsche and these individuals can miss the cross-pollination of ideas. Each person has the same goal: To make the 911 better.

Porsche 911 chief August Achleitner emphasized that he and his colleagues planned every step in the evolution of the 997—from Carrera and Carrera S introduction in 2004 through GT2RS and Turbo S in 2010—in advance, carefully, and logically. Customers and enthusiasts can count on the same "Intelligent Performance" again from Weissach's engineers and Zuffenhausen's management when the next-generation 911 appears. It will be stronger, lighter, safer, and quicker. It will emit less carbon dioxide, slip through the air more efficiently, and require less fuel to achieve greater output. It will look familiar and familial. And it will astonish journalists and customers with its technology and its pure driving pleasure.

YEAR	2009-2010
DESIGNATION	911 Turbo
SPECIFICATIONS	
MODEL AVAILABILITY	Coupe, Cabriolet
WHEELBASE	2350mm/92.5 inches
LENGTH	4477mm/176.3 inches
WIDTH	1852mm/72.9 inches
HEIGHT	1300mm/51.2 inches
WEIGHT	1573kg/3461 pounds
BASE PRICE	Not available
TRACK FRONT	1491mm/58.7 inches
TRACK REAR	1548mm/60.9 inches
WHEELS FRONT	8.5Jx19
WHEELS REAR	11.0Jx19
TIRES FRONT	235/35ZR19
TIRES REAR	305/30ZR19
CONSTRUCTION	Monocoque steel
SUSPENSION FRONT	Porsche Active Stability Management
SUSPENSION REAR	Porsche Active Stability Management
BRAKES	Ventilated, drilled discs, 6-piston front 4-piston rear aluminum monobloc calipers
ENGINE TYPE	Horizontally opposed water-cooled DOHC six-cylinder
ENGINE DISPLACEMENT	3800cc/231.9CID
BORE AND STROKE	102x77.5mm/4.02x3.05 inches
HORSEPOWER	500@6000rpm
TORQUE	480lb-ft@1950-5000rpm
COMPRESSION	9.8:1
FUEL DELIVERY	Direct fuel injection, turbochargers, intercooler
FINAL DRIVE AXLE RATIO	3.44:1
TOP SPEED	194mph

▼ Porsche quoted the top speed at 310 kilometers (194 miles) per hour. The Turbo coupe weighed 1,598 kilograms (3,516 pounds) with PDK. *Photograph Courtesy Porsche Cars North America*

▲ The 530-horsepower Turbo S in coupe or cabriolet form delivered countless Weissach engineering innovations as standard equipment. From PCCB ceramic composite brakes to electronic differential braking with PTV, from twin Variable Geometry Turbine turbochargers to dynamic bending lights, the S was the best of the best. *Porsche Press*

▼ The seven-speed PDK transmission operated through new steering-wheel paddle shifters. Adaptive Sport Seats and Sport Chrono Plus were standard equipment. *Porsche Press*

2010 TURBO, TURBO S, COUPE, CABRIOLET

To answer the call for more power from the Turbo, the company introduced the next-generation Turbo S, which was based on the new 3.8-liter direct fuel-injection engine with twin Variable Geometry Turbine turbochargers. The S version put 530 SAE horsepower under the control of all-wheel-drive technology and transmitted it to the road through the seven-speed PDK with new steering-wheel mounted paddle shifters. This was the gearbox Porsche engineers designed to reliably channel that kind of power output to all four wheels for the life of the car. This fourth-generation S was the best Turbo to that point, not only for its impressive performance, but also for other details, including standard equipment ceramic composite brakes, center-lock wheels, Porsche torque vectoring (which applied differential braking to individual wheels to increase the turning effect), dynamic engine mounts, Sport Chrono Plus, adaptive spot seats, a handsome carbon fiber air intake box, and dynamic bending lights that followed the front steering tires around corners. It was a package—and a packaging masterpiece—that went on to outsell the regular Turbo by almost three to one.

To those most discriminating Porsche customers—those who believe even 250 examples of a 911 Sport Classic represented too much chance of seeing a twin on the road—Porsche Exclusive and American Express teamed up to create the 2010 Porsche 911 Centurion Edition. Inspired by the original 901 shape, the ultra-limited edition celebrated the 10th anniversary of American Express' Black Titanium credit card. The car was based on the Carrera S, but Weissach tweaked Centurion Edition output to 385 SAE horsepower, driven through the PDK. It was the first non-Turbo PDK to offer paddle shifters. Exclusive finished the car in Black Metallic, with a black leather, Alcantara and carbon fiber interior. Centurion car membership required $250,000 in annual expenditures on the car, and it is safe to assume that each of the three U.K. customers who acquired one of these black 911s made a dent in the spending requirement.

As model year 2010 drew to a close, Porsche announced two new models for 2011 that completed the 997 model line: a 911GTS that combined the best of GT3 and Carrera S models, and a 911 Speedster, inspired by the original 356s and carrying on styling cues of the 1989 and 1994 models. One of Porsche's best-managed secrets evolved into a carefully controlled succession of hints about the impending arrival of the next generation 911, which was known as Typ 991 inside Porsche. Development was done, pilot production cars were clad in camouflage, and drivers were out testing the car on public roads as it headed for its debut as a 2012 model.

▲ The GTS designation suggested additional power and improved handling and its roots as a Porsche name went back to its four-cylinder racecars. Weissach engineers adapted some GT3 engine technologies for use with the GTS models. *Porsche Press*

2011-2012 911 CARRERA GTS, GTS CABRIO

With 19 variations already in the 911 vehicle lineup, Porsche's product planners soon found room for another. They took a previously optional "Carrera S Power Pack" and made it the center element of the new GTS model. The 23-horsepower boost from the option came from modifications to cylinder heads, a new intake manifold incorporating a vacuum-control tuning flap for each cylinder rather than one of the entire engine as on the S, and a sport exhaust system. Some of these elements were already part of the 435-horsepower GT3, and the resulting blend placed the GTS—with 408 horsepower—midway in power between it and the Carrera S at 385 horsepower.

The rest of the GTS was a blend as well, making use of rear-wheel drive, but only installing it in the wider, all-wheel drive body. Center locks secured the black RS Spyder wheels. Either the PDK or the six-speed manual transmissions

YEAR	**2011-2012**
DESIGNATION	**Carrera GTS (rear-wheel drive)**
	Carrera 4 GTS (all-wheel drive)
SPECIFICATIONS	
MODEL AVAILABILITY	Coupe, Cabriolet
WHEELBASE	2,450mm/96.5 inches
LENGTH	4,509mm/176.81 inches
WIDTH	1,852mm/72.9 inches
HEIGHT	1,295mm/51.0 inches
WEIGHT	1,425kg/3,142lbs coupe manual
	1,470kg/3,241lbs coupe 4 manual
	1,272kg/3,186lbs coupe PDK
	1,490kg/3,285lbs coupe 4 PDK
	1,450kg/3,296lbs Cabriolet manual
	1,515kg/3,340lbs Cabriolet PDK
	1,540kg/3,395lbs Cabrio 4 manual
	1,560kg/3,439lbs Cabrio 4 PDK
TRACK FRONT	1,532mm/60.31 inches
TRACK REAR	1,518mm/59.76 inches
WHEELS FRONT	9.0Jx20
WHEELS REAR	11.5Jx20
TIRES FRONT	245/35 ZR20
TIRES REAR	305/30 ZR20
CONSTRUCTION	Monocoque Steel/Aluminum/Magnesium
SUSPENSION FRONT	Strut suspension (MacPherson type, Porsche optimized) with wheels independently suspended by transverse link, longitudinal link and struts; cylindrical coil springs with internal dampers, electromechanical power steering
SUSPENSION REAR	Multi-link suspensions with wheels independently suspended on five links; cylindrical coil springs with coaxial internal dampers
BRAKES (Front & Rear)	Six-piston aluminum monobloc front brake calipers; four-piston aluminum monobloc rear brake calipers; perforated and internally ventilated brake discs with 340mm front diameter/330 rear diameter and 34mm front thickness/28mm thickness. Steel rotors
ENGINE TYPE	Horizontally opposed water-cooled DOHC six-cylinder
ENGINE DISPLACEMENT	3,800cc/231.9 CID
BORE AND STROKE	102x77.5mm/4.02x3.05 inches
HORSEPOWER	430@7,500rpm
TORQUE	325lb-ft@5,750rpm
COMPRESSION	12.5:1
FUEL DELIVERY	Direct fuel injection
TOP SPEED	190mph coupe manual
	189mph coupe PDK
	189mph coupe 4 manual
	188mph coupe 4 PDK
	189mph Cabriolet manual
	188mph Cabriolet PDK
	188mph Cabrio 4 manual
	187mph Cabrio 4 PDK

▲ In a tip of its hat to a rapidly emerging market, Porsche unveiled the GT2 RS—its ultimate "street fighter"—at the Moscow auto show. *Porsche Press*

▼ Modifications to the direct fuel injection, to the twin Variable Geometry Turbines and to the intercoolers brought 620 horsepower out of the 3.8-liter flat six. By eliminating anything not necessary for speed and safety, engineers reduced the car's weight to 3,020 pounds. *Porsche Press*

were available. Porsche's Active Suspension Management (PASM) was standard equipment. To satisfy performance addicts, Weissach engineers developed and added a Sport Suspension package, but the vehicle's creator/father, August Achleitner, confided the best choice for all ride and handling situations—except on a track day outing—was to leave the Sport Suspension off. The interior was more GT3 than Carrera S, with generous use of Alcantara throughout. In the U.S., Porsche introduced the coupe at $103,110 and the Cabriolet for $112,900. The company unveiled the car at the October 2010 Paris Auto Show.

2010 911 GT2 RS

Porsche introduced its ultimate version of turbocharged coupe—the 911 GT2 RS—

YEAR	2011-2012
DESIGNATION	Turbo/Turbo S (all-wheel drive)
SPECIFICATIONS	
MODEL AVAILABILITY	Coupe, Cabriolet
WHEELBASE	2350mm/92.5 inches
LENGTH	4477mm/176.3 inches
WIDTH	1852mm/72.9 inches
HEIGHT	1300mm/51.2 inches
WEIGHT	1,570kg/3,461 pounds coupe manual 1,595kg/3,516 pounds coupe PDK 1,608kg/3,544 pounds S coupe PDK 1,645kg/3,627 pounds manual Cabriolet (pdk) 1,670kg/3,682 pounds
TRACK FRONT	1,490mm/58.66 inches
TRACK REAR	1548mm/60.9 inches
WHEELS FRONT	8.5Jx19
WHEELS REAR	11.0Jx19
TIRES FRONT	235/35ZR19
TIRES REAR	305/30ZR19
CONSTRUCTION	Monocoque Steel/Aluminum/Magnesium
SUSPENSION FRONT	Enhanced by PASM with Normal and Sport mode: additionally electrically actuated hydraulic bypass valve for internal damping force.
SUSPENSION REAR	Enhanced by PASM with Normal and Sport mode: additionally electrically actuated hydraulic bypass valve for internal damping force
BRAKES (Front & Rear)	Six-piston aluminum monobloc calipers on front axle; four-piston aluminum calipers on rear axle. Per-axle distribution; internally vented and perforated
BRAKES (Front & Rear)	Turbo S six piston front; four-piston rear Porsche Ceramic Composite Brakes (PCCB)
ENGINE TYPE	Horizontally opposed water-cooled DOHC six-cylinder twin turbochargers, twin intercoolers
ENGINE DISPLACEMENT	3,800cc/231.9 CID
BORE AND STROKE	102 x 77.5mm/4.02 x 3.05 inches
HORSEPOWER	500@6,000rpm Turbo 530@6,250-6,750rpm Turbo S
TORQUE	480lb-ft@1,900-5,000rpm Turbo 516lb-ft@2,100-4,250rpm Turbo S
COMPRESSION	9.8:1
FUEL DELIVERY	Direct fuel injection.
TOP SPEED	194 mph coupe 193 mph Cabriolet 195 mph S coupe

▲ The 997 Speedster provided an updated re-interpretation of Porsche's classic minimalist roadster. The top on the new car was operated by hand, and it attached to a windshield 2.4 inches lower than found on a Carrera cabriolet. *Porsche Press*

at the Moscow International Automobile Salon in August 2010. Starting from the existing GT2, motorsports department engineers pulled out another 154 pounds of material, electronics, and mechanicals to bring the RS down to 3,020 pounds. With special tuning of the DFI flat six, subtle tweaks to the twin Variable Geometry Turbine turbocharger, and slightly modified twin turbochargers, the engine developed 620 horsepower at 6,500 rpm and 516 lb-ft of torque upwards from 2,250 rpm. This provided the rear-drive coupe with staggering performance.

Another rumor suggested that Andreas Preuninger, manager of High Performance Cars, and his crew conceived the RS back in 2007 as a challenge to other manufacturers who were hinting at their own high-horsepower projects. The GT2 RS supposedly had an internal code—"727"—that represented one competitor's best lap time (7 minutes, 27 seconds) around the Nürburgring.

The new RS used Porsche's strong six-speed manual transmission, and the car reached acceleration times from 0-to-60 miles per hour of 3.4 seconds. Ironically, the PDK-equipped Turbo S was 0.3 seconds quicker by virtue of all-wheel-drive traction and computerized shifting. But from there, the RS ran away and on up to a top speed of 205 miles per hour. Porsche set production at only 500 cars, and all were sold within two months. By that time, however, Preuninger's test drivers had sent their own 'Ring record of 7:18.

2010 911 SPEEDSTER

Porsche Exclusive's product planners and parts bin shoppers worked fine magic when creating a new Speedster. For this last celebration of the 997 platform, engineers and stylists mated a version of the recent Sport Classic

YEAR	2011–2012
DESIGNATION	GT2 RS (rear-wheel drive)
SPECIFICATIONS	
MODEL AVAILABILITY	Coupe
WHEELBASE	2350mm/92.5 inches
LENGTH	4,468mm/175.9 inches
WIDTH	1852mm/72.9 inches
HEIGHT	1,285mm/50.6 inches
WEIGHT	1,383kg/3,050 pounds
TRACK FRONT	1,538mm/60.55 inches
TRACK REAR	1,552mm/61.10 inches
WHEELS FRONT	8.5Jx20
WHEELS REAR	11.0Jx20
TIRES FRONT	245/35ZR19
TIRES REAR	305/30ZR20
CONSTRUCTION	Monocoque Steel/Aluminum/Magnesium
SUSPENSION FRONT	Strut suspension (MacPherson type, Porsche optimized) with wheels independently suspended by transverse link, longitudinal link and struts; cylindrical coil springs with internal dampers, electromechanical power steering
SUSPENSION REAR	Multi-link suspensions with wheels independently suspended on five links; cylindrical coil springs with coaxial internal dampers
BRAKES	Front: Six-piston aluminum monobloc brake calipers, perforated and internally ventilated Porsche ceramic composite brakes Rear: Four-piston aluminum monobloc brake calipers, perforated and internally ventilated Porsche ceramic composite brakes
ENGINE TYPE	Horizontally opposed water-cooled DOHC six-cylinder twin variable-geometry turbochargers, twin intercoolers
ENGINE DISPLACEMENT	3,600cc/219.7CID
BORE AND STROKE	100x6.4mm/3.94x3.01 inches
HORSEPOWER	620@6,500rpm
TORQUE	516lb-ft@2,250rpm
COMPRESSION	9.4:1
FUEL DELIVERY	Direct fuel injection
TOP SPEED	(six-speed manual) 206 mph

▶ The Speedster used GTS running gear with its 408-horsepower flat six. But in a major change of character, the only transmission Porsche offered was the paddle-shift automatic PDK transmission. *Porsche Press*

front end to the rear of a Carrera 4S cabriolet body with several distinctive modifications. Both ends and the sides received additional styling treatments. In keeping with Speedster tradition, this new version trimmed 2.4 inches from the windshield height and revised the manually operated convertible roof so it hunkered down through the wind and rain. That top, when collapsed, hid under the revisited twin-bulge first introduced on the 1989 Speedster as the farewell to the G-series platform. The new version debuted alongside the GTS at the Paris Auto Show.

As with the Sport Classic, elements of the GT3 migrated to the new open car as well, including tuning that increased engine output to the same 408 horsepower and 310 lb-ft torque as the GTS. But the Speedster came only with the PDK with paddle-shift. The mix of parts continued with aluminum GT3 doors, but many other Carrera series options were standard gear, including the navigation system; PASM; Sport Chrono Plus; adaptive sport seats; Ceramic composite brakes; and abundant leather on door handles, coat hooks, air vents, and other places. A checkerboard leather motif on the seats was an option buyers could delete at no cost. Exclusive limited production to 356 examples and restricted color choices to Pure Blue or Carrara white.

As *CAR* magazine pointed out, the Speedster was expensive—introduced for 100 U.S. buyers at $204,000—but that steep price covered not only the extra cost options included in the

▲ Porsche enthusiasts who are savvy with company practices have come to expect some of the company's most exciting and desirable offerings just as a model series comes to an end. By lengthening piston stroke, Porsche achieved four-liter displacement for its racing cars, which it detuned for this limited run. *Porsche Press*

standard specification but also the more costly crash testing (different nose on a C4S body) and development of the pop-up rollover bars. As writer Ben Pulman also pointed out, "Part of the package is a louder (but non-switchable) sports exhaust, so you always get to hear the flat six in full war cry mode."

2011 911 GT3 RS 4.0

For many years, Porsche claimed 3.8 liters was the most displacement possible out of their flat six. Such challenges always provoke Andreas Preuninger and his group of GT car engineers. They soon proved a four-liter flat six was possible for competition version GT3 R and RSR models. So, why not offer an end-of-997-run of four-liter cars for their good customers? The GT3 series of cars grew out of homologation needs in the late 1990s, when endurance racing ran under the umbrella of the BPR series, which divided classes into GT1, GT2, and for a short period, GT3 and GT4. Porsche—and all manufacturers competing—needed to assemble a set number of cars to prove their production basis, and Porsche models that previously had been called RS became GT3.

No one was certain how well the cars were going to go over with customers, but marketing hoped for the few hundred to homologate the car for competition. That was about 15 years ago, and, even without offering a GT3 in every single year since, Porsche has sold nearly 15,000 of the GT3 cars. The RS designation returned with a series of the cars aimed very specifically at Porsche track-day enthusiasts, and Preuninger was quick to point out that close to 85 percent of GT3 RS buyers regularly enjoyed their cars on race tracks.

As the 997 series moved to a close, Preuninger's colleagues adopted the longer-stroke engine from the pure competition GT3 models, at 80.4 millimeters instead of the typical 76.4. With titanium connecting rods, aluminum pistons, the crankshaft from the RSR version, and a low-mass flywheel among the hardware changes inside the engine, much of the rest of its enhanced performance came from significantly better airflow by using a revised manifold with shorter runners on the intake side and a lower-back-pressure exhaust. A pair of low-resistance air filters appropriated from the GT3 Hybrid racer further improved breathing. These efforts boosted horsepower from 450 to 494 at 8,250 rpm, and torque from 317 lb-ft up to 339 at 5,750 rpm. Porsche coupled the engine to its well-proven six-speed manual gearbox.

Porsche offered its PCCB carbon composite brake system as an option, and it made dynamic engine mounts standard to stabilize the engine's mass during extremely aggressive driving. To trim a few extra kilograms from the car's weight, the GT car engineers replaced the front deck lid, fenders, and the seats with carbon fiber units. Porsche quoted curb weight at 2,998 pounds—with full fuel tank! For those commuting to their track day outings, the car comes with a sound system and air conditioning, although these are optional deletes. To save another 24 pounds, the standard battery is a lithium-ion battery. Porsche acknowledged that these batteries develop less cranking power in very cold environments, so buyers checking the box for the lightweight battery also found a heavier lead-acid unit in a box for them as well. The company quoted acceleration from 0-to-60 miles per hour in 3.4 seconds with the six-speed manual transmission and a top speed of 193 miles per hour.

Porsche limited production to 600 cars, of which 126 found their way to U.S. buyers. Porsche sold the car for $185,950 in the States.

YEAR	2012
DESIGNATION	**GT3 RS 4.0 (rear-wheel drive)**
SPECIFICATIONS	
MODEL AVAILABILITY	Coupe
WHEELBASE	2,355mm/92.7 inches
LENGTH	4,481mm/176.4 inches
WIDTH	1,808mm/71.2 inches
HEIGHT	1,280mm/50.4 inches
WEIGHT	1,360kg/2,998 pounds
TRACK FRONT	1,538mm/60.55 inches
TRACK REAR	1,552mm/61.10 inches
WHEELS FRONT	9.0x20
WHEELS REAR	12.0Jx20
TIRES FRONT	245/35ZR20
TIRES REAR	325/30ZR20
CONSTRUCTION	Monocoque Steel/Aluminum/Magnesium
SUSPENSION FRONT	Strut suspension (MacPherson type, Porsche optimized) with wheels independently suspended by transverse link, longitudinal link and struts; cylindrical coil springs with internal dampers, electromechanical power steering
SUSPENSION REAR	Multi-link suspensions with wheels independently suspended on five links; cylindrical coil springs with coaxial internal dampers
BRAKES	Front: Six-piston aluminum monobloc brake calipers, perforated and internally ventilated brake discs with 380mm diameter and 34mm thickness Rear: Four-piston aluminum monobloc brake calipers, perforated and internally ventilated brake discs with 350mm diameter and 28mm thickness
ENGINE TYPE	Horizontally opposed water-cooled DOHC six-cylinder
ENGINE DISPLACEMENT	3,996cc/243.8CID
BORE AND STROKE	102x77.5mm/4.02x3.05 inches
HORSEPOWER	500@8,250rpm
TORQUE	339lb-ft@5,750rpm
COMPRESSION	12.5:1
FUEL DELIVERY	Direct fuel injection
TOP SPEED	(manual six-speed) 193 mph

2012–2015 991 CARRERA COUPE, 991 CARRERA S COUPE
2013–2015 991 CARRERA CABRIO, 991 CABRIO S
2013–2015 991 CARRERA 4, 991 CARRERA 4S
2014–2015 991 GT3 (PDK ONLY)
2014–2015 991 TURBO, TURBO S, CABRIO, CABRIO S
2014 991 50TH ANNIVERSARY S
2014–2015 991 TARGA, TARGA 4
2013–2015 991 GT3 RSR
2015 991 GTS, GTS4, GTS CABRIO, GTS4 CABRIO
2015 991 TARGA 4 GTS
2015 991 GTS CLUB COUPE
2016–2018 911 GT3, GT3 RS, R
2017 911 RSR
2017 991/2 CARRERA, CARRERA S
2017 991/2 CARRERA 4, CARRERA 4S
2017 991/2 TARGA 4, TARGA 4S
2017 991/2 TURBO, TURBO S
2017 991/2 911R
2017–2018 911 GTS
2018 TURBO S EXCLUSIVE SERIES
2018 GT3 TOURING
2018 CARRERA T
2018–2020 CARRERA GTS, GTS4
2019 911 GT3 R
2019 935 CLUBSPORT
2019 GT2 RS CLUBSPORT

CHAPTER 9
THE SEVENTH GENERATION 2012–2019

911 TYP 991 CARRERA, CARRERA S

"We had this idea to lengthen the wheelbase long ago, even before 2006, when we first started investigations into the next car," said August Achleitner, the 911 product line director, as he explained one of the controversial dimensions in the 991. "We didn't say exactly 100 millimeters as it is now, but it was roughly this number in 2003. That's when we started development work on the PDK, and we considered the technical requests from our chassis engineers to make the wheelbase longer. It wasn't practical then."

A desire to make the car feel more stable and more precise directed this decision. The wheelbase increase was only part of the effort. Widening the front track was perhaps even more consequential. "Now we can support more rolling forces by the struts, not only by the stabilizer," Achleitner said. "This allowed us to make the stabilizer a little bit thinner and not so stiff. Simultaneously, this avoids under steering. So the new car is much more neutral than the 997. And with this new feature, the PDCC, it's completely another world."

The Porsche Dynamic Chassis Control system first appeared on Porsche's Cayenne and then the Panamera. This variable stabilizer system restrains body roll in aggressive handling situations. It keeps the tires perpendicular to the road in all situations, guaranteeing precise steering and transfer of acceleration or braking loads to the pavement.

New standard tire sizes aid in that task. The 991 Carrera S rides on 20-inch wheels and the Carrera uses 19-inch wheels. "This was one of our target-conflicts," Achleitner said, "because bigger wheels have disadvantages. They need

◄ Longer, lower, wider, faster, more fuel-efficient, yet still a 911 from any perspective. Weissach began developing ideas for the 911 Typ 991 in 2006.

▲ Along with lengthening the wheelbase by 100 millimeters, engineers widened the front track from 1,486 millimeters (58.5 inches), to 1,532 millimeters (60.31 inches). As a consequence, stabilizer diameter decreased, improving ride even as turn-in became more accurate.

▼ The Style Porsche staff characterized this new 911 as the most technical and precise body yet. Until now, the 911 body avoided distinct edges.

more space, they are heavier. One of the best advantages turned out to be that the bigger tire patch let us reduce tire pressure because the potential tire load is higher with these than with smaller tires. It was a little step that not only makes the car faster but also more comfortable."

Beyond performance, driver and passenger comfort have risen ever higher among Weissach's concerns. In the 997, tire, road, and wind noise inside the car was intrusive, and addressing those issues involved chassis and tire engineers, body designers and aerodynamicists, and the interior design team. They went to great lengths to improve the 991.

"By reducing these noises there was some 'space' left in our ears for the engine noise we wanted to hear," Thomas Wasserbäch, general manager for base engine development, said. The result was the Symposer, a joint development with Porsche and the Mann+Kummel Group in neighboring Ludwigsburg. Their multi-chamber module transferred acoustic pulses from the air intake through a funnel-like opening that housed a tuned membrane," Wasserbäch continued. "We developed a new exhaust system that merged the exhaust gas for all cylinders in a new manifold—in the past our Boxer engine sounded like two three-cylinder engines. The new system sounds like there are six cylinders, and we ran a tube from there to the steel bulkhead that became the sub-woofer for our exhaust system. Outside the car you hear exhaust. Inside you hear intake and engine. The sound Symposer brings outside sound in. You can deactivate it if you want a quieter ride by pushing a switch on the console."

The 991 represented a fresh start. That presented innumerable opportunities. Engineers understood there would be 21 variations of this car. Long-term requirements influenced the design and development of the controversial electronic steering.

"We looked at every system out there," Michael Schätzle explained. Schätzle is project manager for the complete 911 product line. "We didn't like anything we drove. So we started from scratch. And that gave us the freedom to develop our own parameters, to address the way Porsche 911 drivers use their cars."

The new system employs a vehicle status sensor that constantly calculates forces impacting the steering rack-and-pinion from road and steering wheel inputs, such as steering angle and vehicle speed. The sensor interprets the data and sends a calculation to an electric motor on the steering rack that applies appropriate steering torque by means of toothed belts and a recirculating ball. Steering feedback, crucial information about road surface and driving conditions, comes to the driver through the steering wheel.

Porsche engaged its longtime development driver Walter Röhrl repeatedly while developing the new car. It was Röhrl who posted a 7:40 Nürburgring lap time that proved every element of the new car's chassis. Nowhere was his input more important than with the new steering system.

"We spent five days developing the steering on the handling course at Nardo in southern Italy," he said. "Then the engineers went back to Stuttgart and worked on parameters. Then we just spent three more days at Nürburgring. That was all it took to make it perfect. One thing the E-steering does is makes driving less fatiguing by filtering out minute road imperfections."

"Our biggest challenge with this new system," Achleitner recalled, returning to the question of long-term development, "was finding a motor that was small enough, light enough, and powerful enough to accommodate forces acting on the front wheels of the 991 GT3 RSR when it bumps the curbing at 200 kilometers an hour.

Porsche offered two versions of E-steering: Power Steering and Power Steering-Plus. Plus provided a little more comfort. Enthusiast drivers preferred the basic version."

The 911 shape is one of the greatest challenges to creating a new model.

"The roofline is inherited from all the 911s that came before," said Matthias Kulla, Porsche general manager for exterior design. "The only thing we can do," Kulla went on, "is to adapt it to the different proportions of the longer wheelbase. We tried to shift the weight point a little further back, not so far forward as the older ones. It starts with the windscreen. It was an aerodynamic consideration to have a 'faster' windscreen. So we left the end where it was, but we pulled the foot of the windscreen forward 70 millimeters, nearly three inches. I think it's the roundest in the industry."

Design chief Michael Mauer challenged his entire staff with an internal competition for concepts for the new 911. "What happens is at first, people have a lot of respect for the car," Kulla explained. "They say, 'Oh, it's the 911. We shouldn't change it.'" He recited the rules: "First, don't be afraid. Be brave. Second thing, don't exaggerate. Third, make it modern."

One of the car's more distinctive new forms is the rear character line. This came from a model that an intern, Peter Varga, created. It was, as Kulla described it, a shape that took the 911 further than anyone imagined while still remaining recognizable as a 911. "The back end character line appeared on a single sketch," Kulla recalled. "And this was where I thought, okay there will be discussions." Another line, an edge that stirred discussions, was the roofline that

▲ A young designer, Peter Varga, introduced these forms to the 911 as part of a scale model he created to receive an internship. These shapes, and other work he did while interning, led to his full-time job.

▲ Porsche body engineers assembled an entire body out of aluminum before concluding there were other ways to save weight. Using a mix of ultra-high-strength steel, aluminum, and magnesium yielded a lighter automobile at lower manufacturing cost.

started at the A-pillar. It accentuated the roof, made it look longer, before finally descending to the rear and putting an elegant sweeping crease around the new, smaller taillights.

"For the longest time, we have had a really soft sloping rear on the 911," he continued. "You probably understand that there are not many areas where we can really make a big step. So it's always in the subtleties. Those are the things that turn a nice car into a really good one. We worked very closely with the engineers and the toolmakers. All these little details have made this by far the most precise 911 ever in terms of execution."

What was the toughest improvement for Achleitner and his engineers to make when advancing from the 997 to the 991?

"For me, it was the reduction of weight," Achleitner admitted. "Because in order to solve this target between pure consumption on one hand and the performance of the car on the other hand, we had to reduce the weight. Otherwise, we would have to install more engine power, and then we cannot reach our targets for consumption.

"In the beginning of the project there were many calculations. Done without having the parts on the table. And out of these calculations, we reached the conclusion—every time we weighed it—that the car was heavier."

"But every engineer and designer had added in a few extra grams," Achleitner said. "Everyone had some small 'reserves.' We put the first car on the weigh station, the first car made from the pre-series where the parts are coming from the final tools . . . and the car was lighter than we calculated. This was a very, very happy moment. And this was without exotic materials. Only with aluminum. Steel. Some magnesium parts."

Lightweight parts fill the updated engines. These were all new for the 997 and substantially changed for the 991. "For the 997 these were the first developments," Thomas Wasserbäch said. "When you work on such a project, you get ideas for the next generation. You write them down, and when that project comes, you know what you want to do. We wanted to bring the engine speed of the GT cars to the serial production; it was our idea to lift the speed 300 rpm more. But to do that you have to make it a little lighter in the drive train."

▲ As it had done while developing the 997, Porsche developed the cabriolet along with the coupe, so introduction of the open car followed within months. The base Carrera offered 350 horsepower while the Carrera S delivered 400. *Porsche Press*

YEAR	2013
DESIGNATION	Carrera (rear-wheel drive)
SPECIFICATIONS	
MODEL AVAILABILITY	Coupe, Cabriolet
WHEELBASE	2,450mm/96.5 inches
LENGTH	4,509mm/176.81 inches
WIDTH	1,852mm/71.18 inches
HEIGHT	1,295mm/51.0 inches
WEIGHT	1,380kg/3,042 pounds coupe 1,450kg/3,197 pounds cabriolet
TRACK FRONT	1,532mm/60.31 inches
TRACK REAR	1,518mm/59.76 inches
WHEELS FRONT	8.5Jx19
WHEELS REAR	11.0Jx19
TIRES FRONT	235/40ZR19
TIRES REAR	285/35ZR19
CONSTRUCTION	Monocoque Steel/Aluminum/Magnesium
SUSPENSION FRONT	Strut suspension (MacPherson type, Porsche optimized) with wheels independently suspended by transverse link, longitudinal link and struts; cylindrical coil springs with internal dampers, electromechanical power steering
SUSPENSION REAR	Multi-link suspensions with wheels independently suspended on five links; cylindrical coil springs with coaxial internal dampers
BRAKES	Four-piston aluminum monobloc brake calipers, perforated and internally ventilated brake discs with 330mm diameter and 28mm thickness. Steel rotors
ENGINE TYPE	Horizontally opposed water-cooled DOHC six-cylinder
ENGINE DISPLACEMENT	3,436cc/209.7CID
BORE AND STROKE	97mmx77.5mm/3.82x3.05 inches
HORSEPOWER	350@7,400rpm
TORQUE	287lb-ft@5,600rpm
COMPRESSION	12.5:1
FUEL DELIVERY	Direct Fuel Injection
TOP SPEED	Coupe 177mph Cabriolet 175mph

The result was better performance—horsepower output, fuel economy, and exhaust emissions—from a slightly reduced displacement. Base 997 engines were 3.6 liters. The 991 base Carrera engine displaced 3.4 liters, but it developed 350 horsepower, five more than the base 997. Porsche offered this engine with its seven-speed PDK or a new seven-speed manual gearbox. The PDK delivered better acceleration, improved fuel economy by 16 percent over the 997, and produced just 205 grams of CO_2 per kilometer driven. Acceleration from 0-to-60 miles per hour took just 4.2 seconds with the PDK and optional Sport Chrono Plus. Top speed was 178 miles per hour.

The Carrera S retained the 3.8 flat six from the 997 but with internal updates and improvements leading to 400 horsepower output, 15 more than the 997 developed. With PDK, fuel economy improved 14 percent over the 997 S, acceleration dropped to 3.9 seconds (with Sport Chrono Plus, and top speed was 187 miles per hour. Auto Stop/Start helped each engine achieve strong fuel economy and emissions improvements. This system worked with either PDK or the manual gearbox. A "sailing" function—essentially dropping the engine to idle speed under no-load conditions—further aided in reducing emissions and fuel consumption.

Within six months of the 991 introduction, Porsche followed with cabriolet bodies on Carrera and Carrera S platforms. The new roofline more closely resembled the coupe profile, and it was the result of hundreds of hours of engineering and design work. While it is longer than the 997 cloth top, the 991's weight is nearly identical. Opening or closing take just 13 seconds, and the top can rise or collapse while the car is moving at speeds up to 31 miles per hour.

YEAR	**2013**
DESIGNATION	**Carrera (rear-wheel drive)**
SPECIFICATIONS	
MODEL AVAILABILITY	Coupe, Cabriolet
WHEELBASE	2,450mm/96.5 inches
LENGTH	4,509mm/176.81 inches
WIDTH	1,852mm/71.18 inches
HEIGHT	1,295mm/51.0 inches
WEIGHT	1,395kg/3,075 pounds coupe 1,465kg/3,230 pounds cabriolet
TRACK FRONT	1,532mm/60.31 inches
TRACK REAR	1,518mm/59.76 inches
WHEELS FRONT	8.5Jx20
WHEELS REAR	11.0Jx20
TIRES FRONT	245/35ZR20
TIRES REAR	295/30ZR20
CONSTRUCTION	Monocoque Steel/Aluminum/Magnesium
SUSPENSION FRONT	Strut suspension (MacPherson type, Porsche optimized) with wheels independently suspended by transverse link, longitudinal link and struts; cylindrical coil springs with internal dampers, electromechanical power steering
SUSPENSION REAR	Multi-link suspensions with wheels independently suspended on five links; cylindrical coil springs with coaxial internal dampers
BRAKES	Front: Six-piston aluminum monobloc brake calipers, perforated and internally ventilated brake discs with 340mm diameter and 34mm thickness Rear: Four-piston aluminum monobloc brake calipers, perforated and internally ventilated brake discs with 330mm diameter and 28mm thickness
ENGINE TYPE	Horizontally opposed water-cooled DOHC six-cylinder
ENGINE DISPLACEMENT	3,800cc/231.9CID
BORE AND STROKE	102mmx77.5mm/4.02x3.05 inches
HORSEPOWER	400@7,400rpm
TORQUE	325lb-ft@5,750rpm
COMPRESSION	12.5:1
FUEL DELIVERY	Direct Fuel Injection
TOP SPEED	Coupe 187mph Cabriolet 185mph

▲ Since introducing the 911 SC Cabriolet as a 1983 model, Porsche's stylists were unhappy with the convertible roof line profile. The 991 soft top matches the contours of the coupe for the first time. *Porsche Press*

▼ The all-wheel-drive 991 incorporated a multi-plate clutch system within the Porsche Traction Management system (PTM) that analyzed and allocated front/rear power distribution ten times per second. Typically, the car operates as a rear-wheel drive vehicle. *Porsche Press*

2013 CARRERA 4, CARRERA 4S FOR COUPES AND CABRIOLETS

While the most significant engineering changes occurred at the front of the Carrera 4 and 4S models, body designer Peter Varga subtly flared the rear fenders 1.7 inches on each side, which was one of the two visual distinctions between rear-drive and all-wheel-drive 991s. The other was the illuminated tail panel that ran below the spoiler, emphasizing the all-wheel drive car's width and tying together the left and rear taillights. As for the engineering, a multi-plate clutch enabled the Porsche Traction Management system (PTM) to alter front/rear power distribution within 100 milliseconds, depending on driving demands and road surface conditions. As Porsche's press release explained, "PTM favors rear-wheel drive when conditions are favorable. Only when adverse slip friction values are sensed (wheel spin, longitudinal acceleration, transverse acceleration, over- or understeer) is torque sent to the front axle in the exact proportion needed to maintain the driver's intended path."

An optional PASM sport chassis lowered ride height by 20 millimeters and incorporated a front spoiler lip unique to Carrera 4 and 4S models. That spoiler lip worked in combination with a higher position for the adaptive rear spoiler to improve 911 aerodynamics. These resulted in zero lift by reducing front lift and increasing rear down force.

Porsche offered its torque vectoring (PTV) system as an option for Carrera 4 models, but it was standard in 4S versions. This system—mechanical with manual gearboxes and electronic with PDK equipped models—precisely decreased the speed of the inside rear wheel during turns to make steering more precise and responsive and to improve acceleration out of turns.

YEAR	2013
DESIGNATION	Carrera 4 (all-wheel drive)
SPECIFICATIONS	
MODEL AVAILABILITY	Coupe, Cabriolet
WHEELBASE	2,450mm/96.5 inches
LENGTH	4,509mm/176.81 inches
WIDTH	1,852mm/71.18 inches
HEIGHT	1,295mm/51.0 inches
WEIGHT	1,430kg/3,153 pounds coupe 1,500kg/3,307 pounds cabriolet
TRACK FRONT	1,532mm/60.31 inches
TRACK REAR	1,560mm/61.42 inches
WHEELS FRONT	8.5Jx20
WHEELS REAR	11.0Jx20
TIRES FRONT	245/35ZR19
TIRES REAR	295/30ZR20
CONSTRUCTION	Monocoque Steel/Aluminum/Magnesium
SUSPENSION FRONT	Strut suspension (MacPherson type, Porsche optimized) with wheels independently suspended by transverse link, longitudinal link and struts; cylindrical coil springs with internal dampers, electromechanical power steering
SUSPENSION REAR	Multi-link suspensions with wheels independently suspended on five links; cylindrical coil springs with coaxial internal dampers
BRAKES (Front & Rear)	Four-piston aluminum monobloc brake calipers, perforated and internally ventilated brake discs with 330mm diameter and 28mm thickness
ENGINE TYPE	Horizontally opposed water-cooled DOHC six-cylinder
ENGINE DISPLACEMENT	3,436cc/209.6CID
BORE AND STROKE	97mmx77.5mm/3.82x3.05 inches
HORSEPOWER	350@7,400rpm
TORQUE	287lb-ft@5,600rpm
COMPRESSION	12.5:1
FUEL DELIVERY	Direct Fuel Injection
TOP SPEED	Coupe 177mph Cabriolet 175mph

YEAR	2013
DESIGNATION	Carrera 4S (all-wheel drive)
SPECIFICATIONS	
MODEL AVAILABILITY	Coupe, Cabriolet
WHEELBASE	2,450mm/96.5 inches
LENGTH	4,509mm/176.81 inches
WIDTH	1,852mm/71.18 inches
HEIGHT	1,295mm/51.0 inches
WEIGHT	1,445kg/3,186 pounds coupe 1,515kg/3,340 pounds cabriolet
TRACK FRONT	1,538mm/60.55 inches
TRACK REAR	1,552mm/61.10 inches
WHEELS FRONT	8.5Jx20
WHEELS REAR	11.0Jx20
TIRES FRONT	245/35ZR19
TIRES REAR	305/30ZR20
CONSTRUCTION	Monocoque Steel/Aluminum/Magnesium
SUSPENSION FRONT	Strut suspension (MacPherson type, Porsche optimized) with wheels independently suspended by transverse link, longitudinal link and struts; cylindrical coil springs with internal dampers, electromechanical power steering
SUSPENSION REAR	Multi-link suspensions with wheels independently suspended on five links; cylindrical coil springs with coaxial internal dampers
BRAKES (Front & Rear)	Front: Six-piston aluminum monobloc brake calipers, perforated and internally ventilated brake discs with 340 mm diameter and 34 mm thickness Rear: Four-piston aluminum monobloc brake calipers, perforated and internally ventilated brake discs with 330mm diameter and 28mm thickness
ENGINE TYPE	Horizontally opposed water-cooled DOHC six-cylinder
ENGINE DISPLACEMENT	3,800cc/231.9CID
BORE AND STROKE	102mmx77.5mm/4.02x3.05 inches
HORSEPOWER	400@7,400rpm
TORQUE	325lb-ft@5,750rpm
COMPRESSION	12.5:1
FUEL DELIVERY	Direct Fuel Injection
TOP SPEED	Coupe 185mph Cabriolet 183mph

▲ The rear fenders flared an additional 1.7 inches per side for the Carrera 4 and 4S bodies. A new taillight panel spanned the rear fascia below the spoiler, visually tying together the left and right taillights while emphasizing the wider body. *Porsche Press*

▶ The latest generation GT3 broke some traditions when project manager Andreas Preuninger specified its only transmission was going to be the seven-speed PDK. He emphasized that for performance drivers interested in speed, the PDK shifted gears faster than they ever could manually. *Porsche Press*

▲ Steering wheel paddle shifts supplemented the center-console-mounted gear selector. Alcantara reduced weight and eliminated distracting reflections. *Porsche Press*

2013 911 TYP 991 GT3 COUPE

In one of its most controversial decisions within the 991 lineup, Porsche introduced its 911 Typ 991 GT3 model only with its seven-speed PDK transmission. The engine—3.8 liters—was tuned to develop 475 horsepower with a 9,000 rpm redline. This was the first GT3 engine with direct fuel injection, using new multi-hole injectors that blasted fuel into the cylinders at as much as 2,900 psi, ensuring exceptional vaporization. This, together with the double-clutch gearbox, provided drivers with 0-to-60 miles per hour acceleration in 3.3 seconds. As GT cars manager Andreas Preuninger explained, "Most of our GT3 customers spend considerable time on the track. For them, the experience is all about speed and being fastest. And, simply, the PDK is the fastest combination. It can shift faster than almost any human." As his release material stated:

"The Porsche dual-clutch transmission in this application has been specially developed for the 911 GT3: the characteristics are based directly on a sequential gearbox from racing, thereby providing further performance and dynamic advantages to the driver. Highlights include shorter gear ratios with closer spacing, even faster shifting, and shift paddles with shorter travel and increased tactile feedback, which allow the driver to place the PDK in neutral simply by pulling on both paddles at the same time."

Porsche also introduced active rear-wheel steering on the GT3. This electronic system enhanced lateral dynamics and steering precision. Below 31 miles per hour, the rear wheels swiveled as much as 2.8 degrees in opposite direction to the fronts. This contracted the turn circle to just 34.8 feet (10.6 meters), which is smaller than any of Porsche's competitors and

YEAR	2013
DESIGNATION	GT3 (rear-wheel drive)
SPECIFICATIONS	
MODEL AVAILABILITY	Coupe
WHEELBASE	2,450mm/96.5 inches
LENGTH	4,509mm/176.81 inches
WIDTH	1,852mm/71.18 inches
HEIGHT	1,269mm/49.96 inches
WEIGHT	1,430kg/3,153 pounds coupe
TRACK FRONT	1,549mm/61.0 inches
TRACK REAR	1,554mm/61.2 inches
WHEELS FRONT	9Jx20
WHEELS REAR	12Jx20
TIRES FRONT	245/35ZR20
TIRES REAR	305/30ZR20
CONSTRUCTION	Monocoque Steel/Aluminum/Magnesium
SUSPENSION FRONT	Strut suspension (MacPherson type, Porsche optimized) with wheels independently suspended by transverse link, longitudinal link and struts; cylindrical coil springs with internal dampers, electromechanical power steering
SUSPENSION REAR	Multi-link suspensions with wheels independently suspended on five links; cylindrical coil springs with coaxial internal dampers
BRAKES (Front & Rear)	Six-piston aluminum monobloc brake calipers, perforated and internally ventilated brake discs with 380 mm diameter and 34mm thickness
ENGINE TYPE	Horizontally opposed water-cooled DOHC six-cylinder
ENGINE DISPLACEMENT	3,800cc/231.9CID
BORE AND STROKE	102mmx77.5mm/231.9 inches
HORSEPOWER	475@8,250 rpm
TORQUE	324lb-ft@6,250rpm
COMPRESSION	12.9:1
FUEL DELIVERY	Direct Fuel Injection
TOP SPEED	195mph

▲ The GT3 introduced all-wheel steering to the Porsche 911 lineup, swiveling rear wheels as much as 2.8 degrees. This improved slow-speed maneuverability and high-speed stability. *Porsche Press*

0.5 meter tighter than the VW Golf. Between 31 and 50 miles per hour, rear steering went along for the ride, but above that the rear wheels pivoted the same direction as the front. They shifted up to 1.8 degrees, creating a crabbing effect that improved agility and lateral stability.

The GT3 used the Carrera 4/4S rear end with 1.7 inches of additional body flare to accommodate wider rear wheels. A distinctive permanent rear wing greatly enhanced airflow over the car. With the new PDK, the GT3 lapped the Nürburgring's Nordschleife in less than 7 minutes 30 seconds.

2014/2015 911 TYP 996 TURBO, TURBO S COUPES AND CABRIOLETS

These cars were the latest interpretations of a technological nexus that Porsche first showed the public in 1973, using engineering that stunned fellow racers back in 1971. In this 2014 version of the 991, the Turbo developed 520

▲ Porsche broke with another 911 tradition when it introduced the Turbo and Turbo S models simultaneously. The S, shown here, offered buyers 560 horsepower, while the Turbo delivered 520. *Porsche Press*

horsepower while the S delivered 560. Torque figures were equally impressive, with the Turbo model providing 487 lb-ft and the S offering 516. For moments when that is simply insufficient, Sport Chrono (standard on the S, optional on the Turbo) "over-boosts" the turbos, yielding 524 lb-ft in the Turbo and 553 in the S for as long as 20 seconds.

Porsche Turbos have run with all-wheel drive since horsepower output topped 400 for the first time with the 911 Typ 933 versions in 1996. A succession of systems has done ever-better jobs of splitting drive. The latest is an electronically activated multi-plate clutch.

"We've had this variable four-wheel-drive clutch, this multi-plate clutch, in the Turbo since the 997 generation," said Erhard Mössle, general manager for Porsche's Turbo, Carrera 4 and Targa 4 lines—all the all-wheel-drive platforms. "But we refined the gearing. Before, we had changes between over- and under-

▲ The decision to standardize all of Porsche's highest-performance 911s with the PDK transmission was based on more than the transmission's ability to handle the load. It allowed Weissach to reduce the number of development prototypes and crash-test cars. *Porsche Press*

steer that was difficult for drivers to see what the car is doing. Now it's clear. And the system works without influencing the driver."

Revising the front-drive involved not only gearing. Mössle's engineers converted the electro-mechanical system to electro-hydraulic. "We deliver more torque to the front axle now because we reinforced the front differential," he said. This inspired them to liquid-cool the front axle. "The higher amount of torque that we can deliver, up to 415 Newton meters [306 lb-ft] to the front axle, means you get a lot of temperature in the system, especially if you are driving with a stronger front axle power ratio," he said.

Directing more than half the engine's output to the front axle requires cooling to ensure reliability. And the 911 remains, by definition, a rear-drive automobile.

Typical driving conditions used the rear axle to manage propulsion duties. Now—for the first time in Porsche's Turbo production history—the rear end contributed directional control not only from the gas pedal but also with new all-wheel steering, first introduced on the GT3. It took the introduction of 991 e-steering to make modern rear steering possible.

"It was a very intelligent solution," Mössle said, "to take one bar from the rear axle and put a motor there to steer the wheels." However, e-steering and all-wheel steering offered Porsche engineers another benefit. "With the 997 four-wheel-drive cars," he continued, "the steering was a bit stiff in the center, not as precise as a two-wheel-drive Carrera. E-steering let us develop a program for the four-wheel-drive cars so they work as well as two-wheel-drive cars. You compensate the higher front axle load with software. That's not possible with hydraulic steering. Now, to the driver, it feels the same."

Porsche's transmission of choice remained the PDK. Porsche coupled this with composite brakes on the Turbo S model. Accommodating rear-end hardware and wider, taller rear tires and wheels (11.5x20 wheels and 305/30 ZR20 Pirelli P-Zeroes) forced Erhard Mössle's engineers to broaden the rear of the car a further 1.1 inches. "Our problem was we got bigger wheels, in diameter and width, and that reduced the space for our charge-air channel going over the wheels. From inlets on the rear fenders going over the wheels we have a plastic tube feeding coolers behind the wheels—a tunnel basically." The larger wheels shrunk the tunnel cross-section, leaving no choice but to widen the car. "That's the reason you have these really brutal, almost flat surfaces going outside the body," he said.

Another new innovation was Porsche Active Aerodynamics (PAA). While rear wings have risen and fallen with speed for more than two decades, the inflatable front air dam is new.

"We discussed it for the 997 Turbo," Mössle recalled, "to bring it in the car, but we would have to do it as an option, so you have two classifications—with and without. This means special chassis set-ups, and you cannot get the cars right with it as an option. And you have a big amount of development cost, so it only works if we put it in the car from the beginning.

"At first we used bicycle tire tubes to push out the spoiler. These were our first trials and errors. Then we found stronger materials, better inflation pumps. So we said when we do systems like rear-wheel steering, adaptive aerodynamics must be standard in all Turbos. And then it filters downstream."

At 75 miles per hour, Speed Mode hoisted the rear wing one inch and inflated two outer segments of the front dam, diverting air around the car body. Performance Mode, activated automatically at speeds of 186 miles per hour (or manually by an instrument panel switch), elevated the rear wing another

▼ The 911 grew wider yet again with the Turbo and Turbo S models. While wide rear wheels had some influence, the major need was to accommodate air intake channels for the twin intercoolers. *Porsche Press*

YEAR	2014
DESIGNATION	Turbo
SPECIFICATIONS	
MODEL AVAILABILITY	Coupe
WHEELBASE	2,450mm/96.5 inches
LENGTH	4,506mm/177.4 inches
WIDTH	1,978mm/77.9 inches
HEIGHT	1,296mm/51.0 inches
WEIGHT	11,595kg/3,516 pounds
TRACK FRONT	1,541mm/60.7 inches
TRACK REAR	1,590mm/62.6 inches
WHEELS FRONT	8.5Jx20
WHEELS REAR	11Jx20
TIRES FRONT	245/35ZR20
TIRES REAR	305/30ZR20
CONSTRUCTION	Monocoque Steel/Aluminum/Magnesium
SUSPENSION FRONT	MacPherson strut aluminum double wishbone
SUSPENSION REAR	Aluminum multi-link
BRAKES	Aluminum monobloc fixed caliper, steel rotors
ENGINE TYPE	Horizontally opposed water-cooled DOHC six-cylinder
ENGINE DISPLACEMENT	3,800cc/231.9CID
BORE AND STROKE	102mmx77.5mm/231.9 inches
HORSEPOWER	520@6,000-6,500rpm
TORQUE	487lb-ft@1,950-5,000rpm
COMPRESSION	9.8:1
FUEL DELIVERY	Direct Fuel Injection. Variable Geometry
TOP SPEED	195mph

YEAR	2015
DESIGNATION	Turbo S
SPECIFICATIONS	
MODEL AVAILABILITY	Coupe
WHEELBASE	2,450mm/96.5 inches
LENGTH	4,506mm/177.4 inches
WIDTH	1,978mm/77.9 inches
HEIGHT	1,296mm/51.0 inches
WEIGHT	1,605kg/3,538 pounds
TRACK FRONT	1,539mm/60.6 inches
TRACK REAR	1,590mm/62.6 inches
WHEELS FRONT	9.0Jx20
WHEELS REAR	11.5Jx20
TIRES FRONT	245/35ZR20
TIRES REAR	305/30ZR20
CONSTRUCTION	Monocoque Steel/Aluminum/Magnesium
SUSPENSION FRONT	MacPherson strut aluminum double wishbone
SUSPENSION REAR	Aluminum multi-link
BRAKES	Aluminum monobloc fixed caliper, composite rotors 410mmx36mm front; 390mmx32mm rear
ENGINE TYPE	Horizontally opposed water-cooled DOHC six-cylinder
ENGINE DISPLACEMENT	3,800cc/231.9CID
BORE AND STROKE	102mmx77.5mm/231.9 inches
HORSEPOWER	560@6,500-6,750rpm
TORQUE	516lb-ft@2,100-4,250rpm
COMPRESSION	9.8:1
FUEL DELIVERY	Direct Fuel Injection. Variable Geometry
TOP SPEED	197mph

YEAR	2015
DESIGNATION	Turbo
SPECIFICATIONS	
MODEL AVAILABILITY	Cabriolet
WHEELBASE	2,450mm/96.5 inches
LENGTH	4,506mm/177.4 inches
WIDTH	1,978mm/77.9 inches
HEIGHT	1,292mm/50.9 inches
WEIGHT	1,665kg/3,671 pounds
TRACK FRONT	1,521mm/60.7 inches
TRACK REAR	1,590mm/62.6 inches
WHEELS FRONT	8.5Jx20
WHEELS REAR	11Jx20
TIRES FRONT	245/35ZR20
TIRES REAR	305/30ZR20
CONSTRUCTION	Monocoque Steel/Aluminum/Magnesium
SUSPENSION FRONT	MacPherson strut aluminum double wishbone
SUSPENSION REAR	Aluminum multi-link
BRAKES	Aluminum monobloc fixed-caliper, steel rotors 380mmx34mm front; 380mmx30mm rear
ENGINE TYPE	Horizontally opposed water-cooled DOHC six-cylinder
ENGINE DISPLACEMENT	3,800cc/231.9CID
BORE AND STROKE	102mmx77.5mm/231.9 inches
HORSEPOWER	520@6,000-6,500rpm
TORQUE	487lb-ft@1,950-5,000rpm
COMPRESSION	9.8:1
FUEL DELIVERY	Direct Fuel Injection. Variable Geometry
TOP SPEED	195mph

two inches and cocked it forward seven degrees. The middle front section inflated and swung forward, joining the two side sections. As this semi-pliable structure diverted more air around the car, it generated a low-pressure area behind it that helped hold down the front end. In this mode—and at 186 miles per hour—these structures generated a combined 291 pounds of down force. Tests on Nürburgring's Nordschleife showed that Performance Mode shortened lap times by two full seconds, taking the Turbo S down to 7:24 on track tires.

The car's new headlamps are another feature destined to filter downstream. Standard on Turbo S models and optional on the Turbo, these LED beams provided light similar in color to daylight, easing eyestrain, improving visual contrast, and reducing driver fatigue. Two vertically stacked tube-shaped light housings provide low- and high-beam illumination. The system modifies beam range based on road speed. The Porsche Dynamic Light System Plus—

YEAR	2015
DESIGNATION	**Turbo S**
SPECIFICATIONS	
MODEL AVAILABILITY	Cabriolet
WHEELBASE	2,450mm/96.5 inches
LENGTH	4,506mm/177.4 inches
WIDTH	1,978mm/77.9 inches
HEIGHT	1,296mm/51.0 inches
WEIGHT	1,595kg/3,516 pounds
TRACK FRONT	1,539mm/60.6 inches
TRACK REAR	1,590mm/62.6 inches
WHEELS FRONT	9Jx20
WHEELS REAR	11.5Jx20
TIRES FRONT	245/35ZR20
TIRES REAR	305/30ZR20
CONSTRUCTION	Monocoque Steel/Aluminum/Magnesium
SUSPENSION FRONT	MacPherson strut aluminum double wishbone
SUSPENSION REAR	Aluminum multi-link
BRAKES	Aluminum monobloc fixed-caliper, composite rotors 410mmx36mm front; 390mmx32mm rear
ENGINE TYPE	Horizontally opposed water-cooled DOHC six-cylinder
ENGINE DISPLACEMENT	3,800cc/231.9CID
BORE AND STROKE	102mmx77.5mm/231.9 inches
HORSEPOWER	560@6,500-6,750rpm
TORQUE	516lb-ft@2,100-4,250rpm
COMPRESSION	9.8:1
FUEL DELIVERY	Direct Fuel Injection. Variable Geometry
TOP SPEED	197mph

optional on GT3 models—replaced the previous Bi-Xenon beams with these LEDs and improved the dynamic cornering function that swiveled headlights toward the inside of a bend based on steering angle and road speed.

Road speed dictated three of Porsche's four specific performance targets for the new Turbo. "We wanted to be under three seconds to 100 kilometers per hour," Mössle says, "and we made 3.1. But, we are always conservative with our figures, so . . . maybe someone beats us. We wanted to be under 10 seconds to 200, and we made 10.3, and we wanted to be under 30 seconds to 300. And we made that in 29.5!

"But our other goal was to be under 10 in fuel consumption." Ten liters per 100 kilometers translates to 23.52 miles per gallon. Ten or above, even in a high-performance car, diminishes some of the car's accomplishment. "Mercedes does 9.9, BMW does 9.9. So we discussed additional features to reach nine-point-something consumption." It also was important to Porsche not to have its U.S. customers pay the Gas-Guzzler tax, measured for 2013 at 22.5 miles per gallon.

Revisions and innovations in several systems achieved the goal. The car's

▲ Porsche introduced new systems on its Turbo models that filtered to other versions in subsequent generations. The PDK provided a "virtual intermediate gear" configuration that better matched road speed to one of the transmission's seven gears by only partially engaging one of the two clutches. *Porsche Press*

active aerodynamics helped, but the 3.3 inches greater width at the rear than the front hurt the effort. One update revised the stop-start system that automatically shuts off the engine after the vehicle comes to a stop. This "driver-defeatable" function restarts the engine as soon as the driver presses the accelerator. This change reduced consumption one tenth of a liter.

New with this Turbo, the PDK's auto-stop function assumed an additional role. The system disengaged its dual clutches while coasting at speed.

▲ For some time, Porsche engineers and stylists had thought about resurrecting the 1960s–1970s treatment for the Targa model. When they introduced the large "panorama" sunroof for the 911, it became clear that this new Targa version was the next logical step. *Porsche Press*

Backing off the throttle down long hills or approaching slower traffic idled the engine while the car coasted in neutral. This same function shut off the engine at slower road speeds when the driver came to a stop.

Still another innovation, confusing on first hearing the terminology, was the "virtual intermediate gears" function on the PDK. While this operated across the entire range of seven gears, its effect was more significant—yet still imperceptible—at lower road speeds. "When you drive in, let's say, second gear at light throttle," Mössle explained, "and the car will shift up into third, the engine may drop to idle speed, too low for driving. So we do not close the clutch completely. We have a little bit of slip in the clutches so you can have revs at, say, about 1,000. So you have a two-and-one-half gear, between second and third."

In the past, Porsche waited a year or longer after Turbo introduction to bring along the S version, "at the end of the car cycle," Mössle said. "With the 997 we did a delay of one year, and that caused big problems with the customers. They bought the 911 Turbo

▲ Once Porsche committed to creating a new, old-style Targa roof, the next big question was whether to operate it electrically or by hand. The push-button choice won out. *Porsche Press*

and as it was delivered, we say, 'Now! Here is the Turbo S'. They told us, 'I would have taken the Turbo S if I had known this.'"

Porsche expected a market to remain for both cars. The S delivered center-lock wheels, ceramic composite brakes, and Sport Chrono among other abilities, items that were options on the Turbo, and for the Turbo cabriolet models, they were features less necessary on an open car. Porsche expected 65 percent of its Turbo customers to order the S.

2015 911 TYP 991 TARGA 4, TARGA 4S

"I wasn't really a big fan of this Targa strategy when we started the long-term planning in 2006 for the sports cars," Grant Larson said. Larson had created the shape and form of the Typ 997, and he was head of exterior design for Porsche's sports cars as the next 911, designated the Typ 991, came into being. When the car's dimensions grew, it offered opportunities to expand the traditional sunroof toward what is called a "panorama" roof by fitting the 997 Targa glass panel instead. That led to discussions about a new version of the Targa. The advanced design and engineering team looked backward into their history books. It occurred to them that updating the original 1960s concept might work.

The 911 Targa, especially since Porsche introduced the cabriolet for 1983, has become an acquired taste. Viewing the car in profile, many Porsche enthusiasts thought the flat, nearly horizontal roof ended abruptly at the brushed stainless steel rollover bar, and it angled too steeply down toward the car's rear deck. "Sometimes different isn't always better," Larson observed. "When we came upon this project, I said, 'we're going to make this roof line perfect.' No compromises—adjust it here or there, back and forth a couple of millimeters, and that's how it came out. The way cars were designed back then, for the mid-sixties, it was brave: Rather than make it black and hide it, we'll make it stainless steel and brush it as bright as possible."

Larson followed the coupe "flow line" for this Targa, and its continuous arc differentiated the new car from the original that debuted in 1965. In profile view, Larson's Targa mirrors the coupe and the new cabriolet.

"Brainstorming in advanced engineering came down to this question," Markus Schulzki said. "Answer yes or no: Fully automatic or manual or . . . ? We talked to our customers. We have many people who say they want a Targa in the old way. But they want to push a button." Schulzki was technical project lead—essentially chief body engineer responsible for making the complicated retractable roof system work by using computer-aided design to accomplish complex kinematics. He and his colleagues considered many ways to open the car and whether they might use a hard top instead of a fabric one. They investigated an integrated linear system,

YEAR	2015
DESIGNATION	**Targa 4 (all-wheel drive)**
SPECIFICATIONS	
MODEL AVAILABILITY	Targa
WHEELBASE	2,450mm/96.5 inches
LENGTH	4,509mm/176.81 inches
WIDTH	1,852mm/71.18 inches
HEIGHT	1,298mm/51.1 inches
WEIGHT	1,540kg/3,395 pounds
TRACK FRONT	1,532mm/60.31 inches
TRACK REAR	1,560mm/61.42 inches
WHEELS FRONT	8.5Jx19
WHEELS REAR	11.0Jx19
TIRES FRONT	245/40ZR19
TIRES REAR	295/35ZR19
CONSTRUCTION	Monocoque Steel/Aluminum/Magnesium
SUSPENSION FRONT	Strut suspension (MacPherson type, Porsche optimized) with wheels independently suspended by transverse link, longitudinal link and struts; cylindrical coil springs with internal dampers, electromechanical power steering
SUSPENSION REAR	Multi-link suspensions with wheels independently suspended on five links; cylindrical coil springs with coaxial internal dampers
BRAKES (Front & rear)	Four-piston aluminum monobloc brake calipers, perforated and internally ventilated brake discs with 330mm diameter and 28mm thickness.
ENGINE TYPE	Horizontally opposed water-cooled DOHC six-cylinder
ENGINE DISPLACEMENT	3,436cc/209.6CID
BORE AND STROKE	97mmx77.5mm/3.82x3.05 inches
HORSEPOWER	350@7,400rpm
TORQUE	287lb-ft@5,600rpm
COMPRESSION	12.5:1
FUEL DELIVERY	Direct Fuel Injection
TOP SPEED	175mph

YEAR	2015
DESIGNATION	**Targa 4S (all-wheel drive)**
SPECIFICATIONS	
MODEL AVAILABILITY	Targa
WHEELBASE	2,450mm/96.5 inches
LENGTH	4,509mm/176.81 inches
WIDTH	1,852mm/71.18 inches
HEIGHT	1,295mm/51.0 inches
WEIGHT	1,445kg/3,186 pounds coupe 1,515kg/3,340 pounds cabriolet
TRACK FRONT	1,538mm/60.55 inches
TRACK REAR	1,552mm/61.10 inches
WHEELS FRONT	8.5Jx20
WHEELS REAR	11.0Jx20
TIRES FRONT	245/35ZR20
TIRES REAR	305/30ZR20
CONSTRUCTION	Monocoque Steel/Aluminum/Magnesium
SUSPENSION FRONT	Strut suspension (MacPherson type, Porsche optimized) with wheels independently suspended by transverse link, longitudinal link and struts; cylindrical coil springs with internal dampers, electromechanical power steering
SUSPENSION REAR	Multi-link suspensions with wheels independently suspended on five links; cylindrical coil springs with coaxial internal dampers
BRAKES	Front: Six-piston aluminum monobloc brake calipers, perforated and internally ventilated brake discs with 340mm diameter and 34mm thickness Rear: Four-piston aluminum monobloc brake calipers, perforated and internally ventilated brake discs with 330mm diameter and 28mm thickness
ENGINE TYPE	Horizontally opposed water-cooled DOHC six-cylinder
ENGINE DISPLACEMENT	3,800cc/231.9 CID
BORE AND STROKE	102mmx77.5mm/4.02x3.05 inches
HORSEPOWER	400@7,400rpm
TORQUE	325lb-ft@5,750rpm
COMPRESSION	12.5:1
FUEL DELIVERY	Direct Fuel Injection
TOP SPEED	183mph

▲ To take advantage of wider tires and more aggressive handling potential, Porsche installed the GTS running gear in the Carrera 4S wide body. GTS ride height with the standard-equipment PASM system is 20 millimeters lower than the Carrera S as well. *Porsche Press*

like the Volkswagen EOS. By 2008, the team knew the roof system was going to be fully automatic.

Schulzki and his engineers worked on the levers and panels that made up the new system and then constructed a scale model to test how the elements moved.

"We made it in one-to-three scale and radio-controlled with the kinematics in scale. You could push a button and the whole system operated," Schulzki said. Unlike the production version, the scale model operated a retractable hardtop. In late 2008, the team demonstrated it to the board, which promptly ordered a full-size prototype. Storing a fixed roof panel compromised rear seating and storage space—challenging the tradi-tional configuration of the 911 as a 2+2. Working from the existing cabriolet folding top gave engineers advantages with the windshield header and integrated locking system. That made the decision to go with a fabric top logical. But no matter how they experimented, removing and storing, the folding top required penetrating the Targa bar that Larson had created in his concepts.

"The way we opened the back part, the window glass," Schulzki explained, "is defined by the way the folding top moved backward. We thought about bringing the rear piece up higher to allow the front folding top to move further under it. But you have much longer levers and more weight. And it didn't look good."

The body engineer's biggest challenge was to develop the mechanism that opened the small flaps on the Targa bow cover without loading additional weight into the car. Like the cabriolet, there was a single electric drive unit, with cylinders for the hinges and the arms to raise the roof system. A push-pull cable system operated the flaps.

This hardware made the Targa 20 kilograms (44 pounds) heavier than the 911 cabrio and added 90 kilograms (198 pounds) to the coupe. To accommodate this, Erhard Mössle and his engineers in the chassis department invented a rebound buffer spring. This miniature coil spring inside the shock absorber piston compensated for the normal abrupt actions of the shock absorber.

Engineers measured the impact forces the suspension transferred from the tires to the glass roof. There were big peaks, so modulating these improved the overall reliability of the glass roof system. For the 997, they incorporated these springs only on the normal suspension. For the 991 platform, they developed it for PASM and fitted it on both front and rear axles.

"We always have discussions," Mössle said, "for the race track with our engineers saying 'go fast, go fast,' and we say, 'think of the customers, the customers.' Not many customers go on the racetrack, but many customers drive the car every day and we tried to come to a compromise between both functions."

Whether Porsche was going to offer a 991 Targa was certain in the earliest advance development in 2006, despite the 997 model's success in selling nearly 8,500 cars on projections of just 7,000 units. When the board members saw Grant Larson's concepts, it struck everyone as a good option.

The entire process requires 19 seconds to open the folding top, fold it on top of itself, hoist it over the Targa bar as the rear window moves up and rearward, settle and stow the Z-folded fabric beneath the rear window, and finally pull back in and reset the large compound back glass panel.

In the process, the back glass extends beyond the furthest rear edge of the car. To clear the trajectory of the fabric panel, it tilts down, obscuring the rear taillights and blocking a clear view of the auxiliary rear stoplight—a situation that violates laws nearly everywhere. Extending the bars to keep the lights visible added length and weight to the steel bars and material costs to the car's price. Allowing operation only at full stop brought the Targa 4 and Targa 4S variations to market underneath the prices of the comparable cabrios.

The 997 Targa outsold company expectations by 20 percent, amounting to about seven percent of all 997 sales. Erhard Mössle described the 991 Targa as offering "coupe-like safety and cabrio-like fun." He said that Porsche believed this new Targa was going to account for 15 percent of sales, cannibalizing some buyers from within the Porsche family—but also stealing others from other makers because of its take on traditional styling and its innovative technology.

▼ Acceleration from 0-to-60 miles per hour took 4.0 seconds with the GTS cabriolet and 3.8 seconds with the coupe when fitted with the optional PDK gearbox. Top speed was 190 miles per hour with the manual transmission. *Porsche Press*

YEAR	2015
DESIGNATION	GTS (rear-wheel drive)
SPECIFICATIONS	
MODEL AVAILABILITY	Coupe, cabriolet
WHEELBASE	2,450mm/96.5 inches
LENGTH	4,509mm/176.81 inches
WIDTH	1,852mm/71.18 inches
HEIGHT	1,295mm/50.98 inches
WEIGHT	1,425kg/3141.58 pounds
TRACK FRONT	1,538mm/66.55 inches
TRACK REAR	1,560mm/61.42 inches
WHEELS FRONT	9Jx20
WHEELS REAR	11.5Jx20
TIRES FRONT	245/35ZR20
TIRES REAR	305/30ZR20
CONSTRUCTION	Monocoque Steel/Aluminum/Magnesium
SUSPENSION FRONT	Strut suspension (MacPherson type, Porsche optimized) with wheels independently suspended by transverse link, longitudinal link and struts; cylindrical coil springs with internal dampers, electromechanical power steering
SUSPENSION REAR	Multi-link suspensions with wheels independently suspended on five links; cylindrical coil springs with coaxial internal dampers
BRAKES	Front: Six-piston aluminum monobloc brake calipers, perforated and internally ventilated brake discs with 340mm diameter and 34mm thickness Rear: Four-piston aluminum monobloc brake calipers, perforated and internally ventilated brake discs with 330mm diameter and 28mm thickness
ENGINE TYPE	Horizontally opposed water-cooled DOHC six-cylinder
ENGINE DISPLACEMENT	3,800cc/231.9CID
BORE AND STROKE	102mmx77.5mm/4.02 x 3.05 inches
HORSEPOWER	430@7,500rpm
TORQUE	325lb-ft@5,750rpm
COMPRESSION	12.5:1
FUEL DELIVERY	Direct Fuel Injection
TOP SPEED	Coupe 190mph Cabriolet 189mph

2015 911 TYP 991 GTS, GTS 4 COUPES AND CABRIOLETS, TARGA 4 GTS

Porsche intended the GTS series to fill a gap between its Carrera S models and the GT3. The result is a blend of engineering and styling features that make this new range perhaps best of both. Weissach engineers developed a new variable resonance induction system that incorporated not only the normal central intake manifold flap (that opens or shuts to vary runner lengths) but also an additional one in each runner synchronized for simultaneous operation determined by engine speed and throttle position, to provide optimal air within the combustion chamber. The result was greater torque at low engine speeds and greater horsepower in higher rev ranges. Coupled with new intake cams, this increased total output to 430 horsepower at 7,500 rpm and raised torque to 325 lb-ft at 5,750 rpm.

Porsche offered GTS buyers their choice of seven-speed manual or optional PDK, but once again the best acceleration came with the PDK, which got the rear-drive coupe from 0-to-60 miles per hour in 3.8 seconds and the cabriolet in 4.0 seconds. In contrast, the highest recorded speed came with the manual transmission rear-drive coupe at 190 miles per hour.

For improved road holding, Porsche selected the Carrera 4S wide body for GTS versions. PASM was standard equipment, as was PTV torque vectoring in manual transmission cars and PTV Plus in PDK-optioned models. As with GT3 and Turbo S models, GTS forged aluminum wheels used center locks. Also similar to the GT3, the GTS version of PASM set standard ride height 20 millimeters below Carrera S models. Porsche Dynamic Chassis Control (PDCC), optional on Carrera and Carrera S models, was optional on GTS variants as well.

YEAR	2015
DESIGNATION	GTS 4 (all-wheel drive)
SPECIFICATIONS	
MODEL AVAILABILITY	Coupe, cabriolet
WHEELBASE	2,450mm/96.5 inches
LENGTH	4,509mm/176.81 inches
WIDTH	1,852mm/71.18 inches
HEIGHT	1,295mm/50.98 inches
WEIGHT	1,470kg/3240.80 pounds coupe 1,540kg/3395.12 pounds cabriolet
TRACK FRONT	1,538mm/60.55 inches
TRACK REAR	1,560mm/61.42inches
WHEELS FRONT	9Jx20
WHEELS REAR	11.5Jx20
TIRES FRONT	245/35ZR20
TIRES REAR	305/30ZR20
CONSTRUCTION	Monocoque Steel/Aluminum/Magnesium
SUSPENSION FRONT	Strut suspension (MacPherson type, Porsche optimized) with wheels independently suspended by transverse link, longitudinal link and struts; cylindrical coil springs with internal dampers, electromechanical power steering
SUSPENSION REAR	Multi-link suspensions with wheels independently suspended on five links; cylindrical coil springs with coaxial internal dampers
BRAKES	Front: Six-piston aluminum monobloc brake calipers, perforated and internally ventilated brake discs with 340mm diameter and 34mm thickness Rear: Four-piston aluminum monobloc brake calipers, perforated and internally ventilated brake discs with 330mm diameter and 2 mm thickness
ENGINE TYPE	Horizontally opposed water-cooled DOHC six-cylinder
ENGINE DISPLACEMENT	3,800cc/231.9CID
BORE AND STROKE	102mmx77.5mm/4.02 x 3.05 inches
HORSEPOWER	430@7,500rpm
TORQUE	325lb-ft@5,750rpm
COMPRESSION	12.5:1
FUEL DELIVERY	Direct Fuel Injection
TOP SPEED	Coupe 189mph Cabriolet 188mph

2015 911 TYP 991 GT3 RS COUPE

At the Geneva International Motor Show in early March 2015, Porsche unveiled its long-anticipated RS version of the 991 GT3 coupe. If there was any surprise for show visitors and journalists, it was the engine: a new 4.0-liter, naturally aspirated flat six. This was not the same four-liter that Porsche enthusiasts had seen a generation before, one derived from the long-lived Hans Mezger designs. This was new. While still developing 500 horsepower, this new engine delivered 338 lb-ft of torque.

As with the 911 Typ 991 GT3, this new car was available only with the strengthened seven-speed PDK transmission. Squeezing both steering-wheel mounted paddles brought this gearbox to neutral, and in a nod to Andreas Preuninger's anticipated customer base for this model, a button on the center console activated a pit lane speed limiter.

As the GT cars engineers had done with the GT3, they again employed the Turbo widebody (and its rear-wheel steering) but with some interesting new enhancements. Most noticeable were vents on the backside of the front wheels that extracted air from the wheel wells. This decreased air pressure under the front of the car, enhancing front downforce. As expected with the RS version of any GT3, the massive fixed (and elevated) rear wing helped push the back end down.

To keep weight low, Porsche used magnesium for the roof panel (for the first time). The roof's wide centerline indentation provided a styling nod to the inset designers and modelers gave to the first-generation 911 front deck lid. The GT3 RS front and rear deck lids were carbon fiber. Alcantara covered most interior surfaces, and the new steering wheel and seats were developed from 918 designs. All this materials management removed 22 pounds of weight from the GT3 RS compared to the GT3. Porsche also announced a Clubsport

version that included a roll cage bolted into the rear of the interior compartment.

Porsche's performance benchmark, a lap of Nürburgring's Nordschleife, required just 7 minutes, 20 seconds. The company quoted acceleration from 0-to-60 miles per hour in 3.1 seconds.

2017 991/2 SECOND-GENERATION CARRERA

There *are* times when less really is more. With steadily increasing pressure on automakers to reduce fuel consumption and exhaust emissions, Porsche's decision to decrease overall displacement to 3.0 liters for its 911 Carrera and Carrera S models makes sense. But this is a Porsche, and Weissach had a goal of introducing performance improvements with each new engine generation. So it became not only sensible but also mandatory that their engineers made twin turbochargers and intercoolers part of the configurations of these updated models.

Some purists moaned over the change and worried about how these new models were to be designated, but the performance results quickly swept aside all concerns. The base Carrera developed 370 horsepower, a 20 horsepower increase over the output of the previous normally aspirated model. For the Carrera S, the improvements were

▲ Porsche fitted 21-inch tires and wheels and at the rear these coped with engine torque and horsepower as well as lateral loads in cornering. Rear-wheel steering aided in making the car feel extremely nimble despite the wide rubber. *Porsche Press*

◀ Creating a Targa 4 GTS was a natural addition to the GTS series. While most Targa owners professed more interest in cruising than aggressive driving, Porsche engineers have understood since the first days of their cars that drivers always want more power! *Porsche Press*

SPECIFICATIONS

YEAR	2017–
DESIGNATION	Carrera Carrera S

MODEL AVAILABILITY	Coupe, cabriolet
WHEELBASE	2,450mm/96.45 inches
LENGTH	4,499mm/177.1 inches
WIDTH	1,808mm/71.2 inches
HEIGHT	1,303mm/51.3 inches
WEIGHT	1,430kg/3,153lbs Carrera manual 1,440kg/3,175lbs Carrera S manual 1,500kg/3,307lbs Cabriolet manual 1,510kg/3,329lbs Cabriolet S manual 1,450kg/3,197lbs Carrera PDK 1,460kg/3,219lbs Carrera S PDK 1,520kg/3,351lbs Cabriolet PDK 1,530kg/3,370lbs Cabriolet S PDK
TRACK FRONT	1,541mm/60.67 inches Carrera 1,543mm/60.74 inches Carrera S
TRACK REAR	1,518mm/59.76 inches
WHEELS FRONT	8.5Jx19 Carrera 8.5Jx20 Carrera S
WHEELS REAR	11.5Jx19 Carrera 11.5Jx20 Carrera S
TIRES FRONT	235/40ZR19 Carrera 245/35ZR20 Carrera S
TIRES REAR	295/35ZR19 Carrera 305/30ZR20 Carrera S
CONSTRUCTION	Monocoque Steel/Aluminum/Magnesium
SUSPENSION FRONT	Porsche-optimized McPherson-type strut suspension with wheels independently suspended by transverse links, longitudinal links, and struts; cylindrical coil springs with internal dampers; electromechanical power steering; optional front lift system
SUSPENSION REAR	Multi-link suspension with wheels independently suspended on five links; cylindrical coil springs with coaxial internal dampers
OTHER	Porsche Active Suspension Management (PASM); with electronically controlled dampers, two manually selectable damping programs
BRAKES (Front & Rear)	Front: Four-piston aluminum monobloc brake calipers, perforated and internally ventilated brake discs of 330mm diameter and 34mm thickness Rear: Four-piston aluminum monobloc brake calipers, perforated and internally ventilated brake discs of 330mm diameter with 28mm thickness
ENGINE TYPE	Horizontally opposed water-cooled DOHC six-cylinder
ENGINE DISPLACEMENT	2,981cc/181.9CID
BORE AND STROKE	91.0x76.4mm/3.58x3.01 inches
HORSEPOWER	370@6,500 rpm Carrera 420@6,500 rpm Carrera S
TORQUE	331 lb-ft@1,700–5,000 rpm Carrera 368 lb-ft@1,700–5,000 rpm Carrera S
COMPRESSION	10:1
FUEL DELIVERY	Direct Fuel Injection, twin turbochargers, twin intercoolers
TRANSMISSION	Standard: Seven-speed manual transmission with two-plate clutch Optional: Seven-speed dual-clutch transmission (PDK)
TOP SPEED	Cabriolet manual 181mph Cabriolet S manual 190mph Carrera PDK 182mph Carrera S PDK 190mph Cabriolet PDK 180mph Cabriolet S PDK 189mph

similar, with horsepower up to 420 (also a 20 horsepower gain). While each model uses the same 3.0-liter displacement engine, Weissach engineers modified the turbo compressors, changed the exhaust system, and revised engine management electronics to achieve the 50 horsepower difference between the base and S versions.

Increased torque was an expected byproduct of these engineering changes. The base car developed 331 lb-ft of torque and the S produced 368 lb-ft, both delivering this performance across an engine rev band from 1,700 up to 5,000 rpm. These new engines had a maximum revolution speed of 7,500 rpm, unusually high for a turbocharged engine and one that coincidentally guaranteed that these engines produced the much-loved Porsche sound despite the additional sound deadening that turbocharging induces. For the Porsche engineers involved in this new engine project, the decision came not only from international regulations but also from personal pride.

"If you drive in Germany," Thomas Krickelberg said, "on country roads or the autobahn with our normally aspirated engines, they are fascinating." Krickelberg was overall project manager for the second-generation 991 cars. "Their sound is different. But all the other cars are turbocharged, they have strong diesel engines, or they are super-charged. In a normally aspirated 911, you had to push them really hard to get some distance; you had to rev your engine high to keep them away. And that's hard work. But now with these turbo engines, it's gotten really much easier."

Getting to those autobahn speeds was much easier too, with 0-to-100 kilometer-per-hour acceleration trimmed by two-tenths of a second. Acceleration in the Carrera with PDK transmission and the Sport Chrono package took 4.2 seconds, and for the Carrera S, that time span dropped to 3.9 seconds, marking the first time a 911 Carrera undercut that significant 4-second elapsed time. For perspective, 3.9 seconds was exactly the time it took the legendary twin turbo 959 to reach 100 kilometers per hour. The Carrera S top speed of 192 miles per hour is just 4 miles per hour short of the 959's terminal velocity.

To keep up with the more powerful engine, Weissach engineers developed a new Porsche Active Suspension Management (PASM) chassis that lowered ride height by 0.39 inches (10mm) to improve handling and stability in cornering. Next-generation shock absorbers contributed to this stability even as they improved ride comfort. Active rear-wheel steering, introduced on the 991 Turbo and GT3, was offered as an option on the 991/2 Carrera S models. The new steering wheel was an adaptation of what appeared on the 918 Spyder.

▲ Porsche debuted the second-generation 991-series 911 at the Frankfurt International Auto Show in September 2015. The list of "what's new" included—most controversially—twin turbochargers on models called the Carrera and Carrera S. *Porsche Press*

▼ The most distinctive visual change—accomplished with typical Style Porsche subtlety—was the addition of intercooler exit vents at the rear of the car to the side of the exhausts. This Carrera boasted 370 horsepower while the Carrera S provided 420 horsepower. *Porsche Press*

▶ While the new Carrera's interior looked familiar, there were many changes and updates. Most significant was the new Porsche Communication Management (PCM) system that included an online navigation module and voice control on its seven-inch screen. *Porsche Press*

▼ Porsche introduced the 2017 all-wheel-drive Carrera 4 and 4S models at the Tokyo Motor Show at the end of October 2015. The same distinctive back-end sculpting carried over from the rear-wheel drive models and incorporated exit vents behind the rear wheels for intercooler heat. *Porsche Press*

YEAR	2017-
DESIGNATION	Carrera 4, Carrera 4S, Targa 4, Targa 4S

SPECIFICATIONS

MODEL AVAILABILITY	Coupe, Cabriolet
WHEELBASE	2,450mm/96.45 inches
LENGTH	4,499mm/177.1 inches
WIDTH	1,852mm/72.9 inches
HEIGHT	
1,294mm/50.9in	Carrera 4 coupe
1,296mm/51.0in	Carrera 4S coupe
1,289mm/50.7in	Carrera 4 cabriolet
1,291mm/50.8 in	Carrera 4S cabriolet
1,288mm/50.7in	Targa 4
1,293mm/50.9in	Targa 4S
WEIGHT	
1,430kg/3,263lb	Carrera 4 manual
1,440kg/3,285lb	Carrera 4S manual
1,500kg/3,417lb	Cabriolet 4 manual
1,510kg/3,439lb	Cabriolet 4S manual
1,450kg/3,307lb	Carrera 4 PDK
1,460kg/3,329lb	Carrera 4S PDK
1,520kg/3,461lb	Cabriolet 4 PDK
1,530kg/3,483lb	Cabriolet 4S PDK
1,570kg/3,461lb	Targa 4 manual
1,580kg/3,483lb	Targa 4S manual
1,590kg/3,505lb	Targa 4 PDK
1,600kg/3,527lb	Targa 4S PDK
TRACK FRONT	1,541mm/60.67 inches Carrera 4 Coupe, Cabriolet, Targa 1,543mm/60.74 inches Carrera 4S Coupe, Cabriolet, Targa
TRACK REAR	1,558mm/61.34 inches
WHEELS FRONT	8.5Jx19 Carrera 4, Targa 4 8.5Jx20 Carrera 4S, Targa 4S
WHEELS REAR	11.5Jx19 Carrera 4, Targa 4 11.5Jx20 Carrera 4S, Targa 4S
TIRES FRONT	235/40ZR19 Carrera 4, Targa 4 245/35ZR20 Carrera 4S, Targa 4S
TIRES REAR	295/35ZR19 Carrera 4, Targa 4 305/30ZR20 Carrera 4S, Targa 4S
CONSTRUCTION	Monocoque Steel/Aluminum/Magnesium
SUSPENSION FRONT	Porsche-optimized McPherson-type strut suspension with wheels independently suspended by transverse links, longitudinal links, and struts; cylindrical coil springs with internal dampers; electromechanical power steering; optional front lift system
SUSPENSION REAR	Multi-link suspension with wheels independently suspended on five links; cylindrical coil springs with coaxial internal dampers
OTHER	Porsche Active Suspension Management (PASM); with electronically controlled dampers, two manually selectable damping programs
BRAKES (Front & Rear)	Front (Carrera 4): Four-piston aluminum monobloc brake calipers, perforated and internally ventilated brake discs of 330mm diameter and 34mm thickness Front (Carrera 4S): Six-piston aluminum monobloc brake calipers, perforated and internally ventilated brake discs of 350mm diameter and 34mm thickness Rear: Four-piston aluminum monobloc brake calipers, perforated and internally ventilated brake discs of 330mm diameter with 28mm thickness
ENGINE TYPE	Horizontally opposed water-cooled DOHC six-cylinder
ENGINE DISPLACEMENT	2,981cc/181.9CID
BORE AND STROKE	91.0x76.4mm/3.58x3.01 inches
HORSEPOWER	370@6,500rpm Carrera 420@6,500rpm Carrera S
TORQUE	331 lb-ft@1,700–5,000 rpm Carrera 368 lb-ft@1,700–5,000rpm Carrera S
COMPRESSION	10:1
FUEL DELIVERY	Direct Fuel Injection, twin turbochargers, twin intercoolers
TRANSMISSION	Standard: Seven-speed manual transmission with two-plate clutch, active all-wheel drive Optional: Seven-speed dual-clutch transmission (PDK) with active all-wheel drive
TOP SPEED	Carrera 4 manual 181mph Carrera 4S manual 189mph Cabriolet 4 manual 179mph Cabriolet 4S manual 188mph Targa 4 manual 180mph Targa 4S manual 188mph Carrera 4 PDK 180mph Cabriolet 4 PDK 178mph Targa 4 PDK 178mph Carrera 4S PDK 188mph Cabriolet 4S PDK 187mph Targa 4S PDK 187mph

Weissach electronics engineers completely revised the Porsche Communications Management System (PCM) with voice control and online navigation. Even handwritten inputs were possible on the new seven-inch display screen. Mobile phones connected by WiFi; a smartphone tray integrated into the center console charged the battery and maximized cellular phone reception. Electronics benefited typical driving situations as well; optional Adaptive Cruise Control (ACC) gently applied the brakes to long downhill runs when ground speed passed the pre-set, and it offered a coasting function for cars equipped with the PDK.

Body styling changes were understated, with Matthias Kulla's exterior design team dedicated to keeping the distinction clear between Carrera models and the car called the Turbo. "My goal," he said, "was always to have the base-model 911 to be the purest. . . . There was one area where someone could recognize that the new 911 had a turbocharged engine, and it's the air outlet for the intercoolers. Apart from that, it's very hard to tell."

Porsche debuted the second-generation 991 in September 2015 at the IAA in Frankfurt to a large, tightly packed crowd of enthusiasts and journalists. The U.S. unveiling took place at the Los Angeles show in November. In typical form, cabriolet models on Carrera and Carrera S chassis followed quickly, as did Carrera 4 and 4S all-wheel-drive versions. Then came Targa 4 and Targa 4S variations, as well as the GTS-configured models on rear- and all-wheel-drive platforms.

2017 991/2 TURBO AND TURBO S

Weissach's Turbo engineers modified engine inlet ports, changed fuel injection nozzles, and increased the fuel

▶ The all-new twin-turbocharged 911 engine displaced 2,981cc with bore and stroke of 91.0 by 76.4mm. This S engine developed 420 horsepower at 6,500 rpm. Engineers adopted turbos and intercoolers in order to meet increasingly strict CO_2 emissions and fuel economy standards while still providing improved performance over the previous generation's 991 models. *Porsche Press*

▶ Porsche returned to a familiar venue for its open cars—the Los Angeles International Auto Show in mid-November 2015—to unveil the next-generation Targa. The new turbocharged engines provided more power and better fuel economy to this distinctive model. *Porsche Press*

▼ There was nothing simple about the new second-generation 991 models. More demanding fuel economy and exhaust emission standards and ever strengthening vehicle safety requirements gave engineers (whose goal always was better performance, handling, and comfort) a nearly infinite list of challenges. *Porsche Press*

▲ Porsche continued circling the globe with model introductions, using Detroit's North American International Auto Show in mid-January 2016 as the launch point for the second-generation Turbo and Turbo S models. Longitudinal grilles and a center louvered engine air intake marked some of the most distinctive appearance changes to this model. *Porsche Press*

injection pressure to increase output by 20 horsepower for both their Turbo and Turbo S models, taking them to 540 and 580, respectively. The S engine also received new variable turbine geometry (TVG) turbochargers using larger compressors than those on the base Turbo model. In addition, engine electronics functions created what Porsche called a dynamic boost function that—using the VTG turbos—maintained charge pressure even when the driver released the throttle for an instant. This system kept the throttle valve open even while interrupting fuel feed. The result in driving dynamics was virtually no turbo boost delay.

These revised engines delivered stunning acceleration. The Turbo coupe raced from 0 to 100 kilometers per hour in 3.0 seconds and topped out at 199 miles per hour. The Turbo S coupe streaked from 0 to 100 in 2.9 seconds, reaching a top speed of 205 miles per hour. This performance last was seen on the lethally potent GT2 RS model, a rear-wheel-drive 911 known for requiring exceptional driving skill. The all-wheel drive Turbo and Turbo S achieved similar performance with far kinder driving manners. These behavioral manners relied on contributions from the standard-equipment Porsche Dynamic Chassis Control (PDCC) and Porsche Ceramic Composite Brake (PCCB) system on the Turbo and Turbo S models. In addition, the new Turbos offered an optional front-lift system that elevated the nose of the car 1.57 inches for steep driveway approach-angle clearance.

The new Turbo and Turbo S coupes and cabriolets adopted the 918-style

YEAR	2017-
DESIGNATION	Turbo Turbo S

SPECIFICATIONS

MODEL AVAILABILITY	Coupe, cabriolet
WHEELBASE	2,450mm/96.45 inches
LENGTH	4,507mm/177.4 inches
WIDTH	1,880mm/74.0 inches
HEIGHT	1,297mm/51.0 inches Turbo, Turbo S Coupe 1,294mm/50.9 inches Turbo, Turbo S Cabriolet
WEIGHT	1,595kg/3,516 pounds Turbo Coupe 1,665kg/3,670 pounds Turbo Cabriolet 1,600kg/3,527 pounds Turbo S Coupe 1,670kg/3,682 pounds Turbo S Cabriolet
TRACK FRONT	1,541mm/60.74 inches
TRACK REAR	1,590mm/59.76 inches
WHEELS FRONT	9Jx20
WHEELS REAR	11.5Jx20
TIRES FRONT	245/35ZR20
TIRES REAR	305/30ZR20
CONSTRUCTION	Monocoque Steel/Aluminum/Magnesium
SUSPENSION FRONT	Porsche-optimized McPherson-type strut suspension with wheels independently suspended by transverse links, longitudinal links, and struts; cylindrical coil springs with internal dampers; electromechanical power steering; optional front lift system
SUSPENSION REAR	Multi-link suspension with wheels independently suspended on five links; cylindrical coil springs with coaxial internal dampers; active real wheel steering
OTHER	Porsche Active Suspension Management (PASM); with electronically controlled dampers, two manually selectable damping programs
BRAKES (Front & Rear)	Front (Turbo): Six-piston aluminum monobloc brake calipers, perforated and internally ventilated ceramic brake discs of 380mm diameter and 34mm thickness Front (Turbo S): Six-piston aluminum monobloc brake calipers, perforated and internally ventilated ceramic brake discs of 410mm diameter and 36mm thickness Rear (Turbo): Six-piston aluminum monobloc brake calipers, perforated and internally ventilated brake discs of 380mm diameter with 30mm thickness Rear (Turbo S): Six-piston aluminum monobloc brake calipers, perforated and internally ventilated brake discs of 390mm diameter with 32mm thickness
ENGINE TYPE	Horizontally opposed water-cooled DOHC six-cylinder
ENGINE DISPLACEMENT	3,800cc/231.9CID
BORE AND STROKE	102.0x77.5mm/4.09x3.05 inches
HORSEPOWER	540@6,400rpm Turbo 580@6,750rpm Turbo S
TORQUE	524lb-ft@2,250–4,000rpm Turbo 553lb-ft@1,700–5,000rpm Turbo S
COMPRESSION	9.8:1
FUEL DELIVERY	Direct Fuel Injection, twin Variable Turbine Geometry turbochargers, twin intercoolers
TRANSMISSION	Standard: Seven-speed dual-clutch transmission (PDK) with active all-wheel drive
TOP SPEED	Turbo 199mph Turbo S 205mph

steering wheel introduced on the second-generation 991 Carrera and Carrera S models. The same PCM improvements moved into the next generation Turbo models. As with first-generation 991 Turbos, Porsche has carried over the Bose sound system as standard equipment with a Burmester system as an optional upgrade.

The exterior of the Turbo and Turbo S enjoyed the same modest facelift as the Carrera models, incorporating the four-LED running lights and introducing dramatic side airblades with additional LEDs that flanked the central front air intake. For the first time, Turbo and Turbo S versions ran with 20-inch tires front and rear (though of different widths.) The rear of the car, representing the business end of any Porsche 911 named "Turbo," became more dramatic as well, with longitudinal louvers on the rear deck lid (a first since prototypes in the early 1960s) and a separate intake for engine air.

In the United States, Porsche Cars North America debuted the new Turbo and Turbo S models at the North American International Auto Show in Detroit in January 2016.

2017 911R

No other car manufacturer has so completely mastered the fine art of conceiving and marketing special editions of their main line production. It started with the Porsche 911 Carrera 2.7 RS. That model designation proved the truth of the phrase "If you build it, they will come." The motivation for mass-producing this car was to homologate it for racing. Unveiled in 1972 and offered as a 1973 model, the RS 2.7 was exclusive, and as such, it became very appealing, especially to those in nations where Porsche chose to not sell it. The speed with which it sold out caught

◀ The new 360mm GT sport steering wheel incorporated a mode switch and Sport Response button in the Turbo. The switch rotated to select Normal, Sport, Sport Plus, or Individual settings in the standard-equipment Sport Chrono Plus package. *Porsche Press*

▼ Both Turbo and Turbo S models took pride in being the only automobiles in series production to use Variable Geometry Turbines (VGT). Two slightly different, very complicated, and highly efficient systems boosted the 3.8-liter displacement Turbo to 540 horsepower output and 580 for the Turbo S. *Porsche Press*

▲ Porsche chose the Geneva Motor Show—the site of many of the company's past racing car debuts—to introduce the new 911R as what is likely to be the last normally aspirated 911. Using the four-liter flat-six carried over from the GT3 RS 4.0, the 3,020-pound car responded well to the engine's 500 horsepower output. *Porsche Press*

▼ The original 911R models of 1967/1968 proved their capability during many long-distance events. When Style Porsche created the color scheme for the new R, they chose to honor a 96-hour record run in an R done at Monza by four Swiss drivers whose national racing colors were red and white. *Porsche Press*

YEAR	2016, 2017-
DESIGNATION	911 GT3 GT3 RS, R

SPECIFICATIONS

MODEL AVAILABILITY	Coupe
WHEELBASE	2,457mm/96.73 inches
LENGTH	4,545mm/178.9 inches
WIDTH	1,852mm/72.9 inches
HEIGHT	1,269mm/49.9 inches GT3, R 1,29mm/50.8 inches GT3 RS
WEIGHT	1,430kg/3,153 pounds GT3 1,420kg/3,131 pounds GT3 RS 1,370kg/3,020 pounds R
TRACK FRONT	1,551mm/61.1 inches GT3 1,587mm/62.5 inches GT3 RS, R
TRACK REAR	1,555mm/61.2 inches GT3 1,557mm/61.3 inches GT3 RS, R
WHEELS FRONT	9Jx20 GT3 9.5Jx20 GT3 RS, R
WHEELS REAR	12Jx20 GT3 12.5Jx21 GT3 RS, R
TIRES FRONT	245/35ZR20 GT3 265/35ZR20 GT3 RS, R
TIRES REAR	305/30ZR20 GT3 325/30ZR21 GT3 RS, R
CONSTRUCTION	Monocoque Steel/Aluminum/Magnesium
SUSPENSION FRONT	Porsche-optimized McPherson-type strut suspension with wheels independently suspended by transverse links, longitudinal links, and struts; cylindrical coil springs with internal dampers; electromechanical power steering; optional front lift system
SUSPENSION REAR	Multi-link suspension with wheels independently suspended on five links; cylindrical coil springs with coaxial internal dampers; active rear-wheel steering
OTHER	Porsche Active Suspension Management (PASM); with electronically controlled dampers, two manually selectable damping programs
BRAKES (Front & Rear)	Front (GT3, GT3 RS): Six-piston aluminum monobloc brake calipers, perforated and internally ventilated ceramic brake discs of 380mm diameter and 34mm thickness Front (R): Six-piston aluminum monobloc brake calipers, perforated and internally ventilated ceramic brake discs of 410mm diameter and 36mm thickness Rear (GT3, GT3 RS): Six-piston aluminum monobloc brake calipers, perforated and internally ventilated ceramic brake discs of 380mm diameter with 30mm thickness Rear (R): Six-piston aluminum monobloc brake calipers, perforated and internally ventilated ceramic brake discs of 390mm diameter with 32mm thickness
ENGINE TYPE	Horizontally opposed water-cooled DOHC six-cylinder
ENGINE DISPLACEMENT	GT3 3,799cc/231.9CID GT3 RS, R 3,996cc/243.9CID
BORE AND STROKE	GT3 102.0x77.5mm/4.09x3.05 inches GT3 RS, R 102.0x81.5mm/4.09x3.21 inches
HORSEPOWER	GT3 475@8,250rpm GT3RS 500@8,250rpm R 500@8,250rpm
TORQUE	GT3 325 lb-ft@6,250rpm GT3 RS, R 339lb-ft@6,250rpm
COMPRESSION	12.9:1
FUEL DELIVERY	Direct Fuel Injection
TRANSMISSION	GT3, GT3 RS Seven-speed DPK R Six-speed manual
TOP SPEED	GT3 196mph GT3 RS 193mph R 201mph

some in the company by surprise, and it taught others the lesson of offering customers something special.

They had passed on a similar opportunity five years earlier with a car called the 911R. Porsche introduced this model at Hockenheimring to journalists in the fall of 1967 as a 1968 model. It was a no-compromises, race-bred-and-racing-purposes-built 911, and although marketing and press director Huschke von Hanstein urged Porsche to manufacture a run of 500 or more, no one else thought the idea worthwhile. That included the car's inventor Ferdinand Piëch, who wanted the vehicle strictly for experiments and competition.

Piëch created four prototypes, according to factory records, and then had the experimental department "manufacture" 20 more as "production" cars. And then it was done. The R took on mythological proportions when one of the fleet set 17 world and international speed records over 96 hours in flat-out laps around Italy's Monza circuit. This happened in a car with an engine that previously had completed a 100-hour bench test, had been taken apart for examination, and re-assembled, though not rebuilt. The car ran with twin longitudinal red stripes to celebrate the Swiss nationality of its four drivers. Other Rs won the Marathon de la Route and their class in the Tour de France Tour de France d'Auto.

The Rs excelled at long-distance punishment. Still, the car was not for everyone. The last two weren't even delivered to their first owners until sometime in 1970! That fact convinced Porsche marketing that von Hanstein was crazy thinking this Spartan 911 would sell. When the luxury version of the 1973 Carrera 2.7 RS—called "Touring"—helped generate sales of more than 1,500 units, some of the

▲ If you are a serious Porsche performance enthusiast, this was the car you wanted most, and this was the interior you wanted most to occupy. Porsche limited production to 991 examples of this 6-speed manual transmission–equipped coupe. *Porsche Press*

earlier skeptics began to think that von Hanstein's idea was crazy like a fox.

With literally dozens of special models passing through Zuffenhausen assembly doors since the 1970s, it was little surprise that, commemorating the 1967 debut of the original R, Porsche resurrected the model designation once again to celebrate the last of the normally aspirated 911s. And they chose, in most of its promotion, to honor the Monza record run history, showing the car in basic white with the Swiss driver red stripes. The promo photos that appeared with green stripes seemed to acknowledge a period in time when past owners, current enthusiasts, and others trimmed the record-run car in the colors of its most important sponsor, BP. Porsche, an equal opportunity historian, honored both versions of the amazing story.

The new R was based on the first-generation 991 GT3 RS 4.0 and used the same four-liter, water-cooled flat-six developing 500 horsepower at 8,500 rpm. Honoring the traditions set by the first R models of 1967 and 1968, this new car was available only with a six-speed manual transmission with a shortened gearshift lever and an optional single-mass flywheel. Acceleration from 0 to 100 kilometers per hour took 3.8 seconds, and Porsche quoted a top speed of 201 miles per hour.

As the original R was a pure, simple, purpose-built performance machine, Porsche provided the new namesake with similar tricks and traits—and several new ones. Rear-axle steering was specially retuned for this version, and the ceramic composite brakes (PCCB) were standard with 410mm front rotors and 390s on the rear.

One of the primary purposes of the original R was to develop the lightest weight 911 for competition purposes. Weissach engineers managed to trim the weight of the new car to 3,020 pounds, 110 pounds less than the already-slimmed GT3 RS. They accomplished this by adopting the material mix of the RS 4.0 with carbon fiber fenders and deck lid, and the magnesium roof panel. Reducing interior insulation and deleting the rear seat helped further as did rendering the air conditioning and audio systems as options rather than standard equipment.

The rear view of the car presented the clean appearance of the standard Carrera (though with a more serious pair of exhaust pipes at the center.) The original R predated the 1973 Carrera 2.7 RS with its ducktail by five years, and without that aerodynamic aid (and with the original car's shorter wheelbase) driving the car fast required tremendous talent to accommodate its twitchy agility and rear lift. Neither of those characteristics were ones Porsche wanted to revisit, so, in the case of air flow management, the new R uses the retractable spoiler from the Carrera models.

Porsche unveiled the 2017 911 R at the Geneva Motor Show in early March 2016. The company limited production to 991 examples.

2016 911 TARGA 4S EXCLUSIVE DESIGN EDITION

Porsche took only a few months after announcing the 911 R before introducing its next "retro" 911, its Targa 4S Exclusive version, as a "collector's item" painted in Etna Blue. This was a standard color from the early 1960s color charts for the Typ 356 B (T5). The

▲ The 2016 911 Targa 4S Exclusive Edition, in Aetna Blue. *Porsche Presse*

Style Porsche designers working for Exclusive finished many details of the 420-horsepower 911 in body color while others appeared in white-gold metallic with a satin finish. Porsche unveiled it at the mid-August AvD Oldtimer Grand Prix at the Nürburgring.

PORSCHE EXPERIENCE CENTER LOS ANGELES

On November 15, 2016, the eve of the Los Angeles International Auto Show, Porsche opened its third Porsche Experience Center (PEC) in Carson, California, just south of Los Angeles International Airport. The $60 million facility is similar to the one at Porsche Cars North America headquarters in Atlanta, Porsche's second PEC and first in the United States. The centers offer potential customers several programs to drive cars with instruction—a much more effective test drive than what is available on public streets. Like Atlanta, PEC LA has a full-service restaurant overlooking the track complex. Significantly, the structure's size brought the Porsche Motor Sports North American (PMNA) operation onsite with large glass walls that allow visitors to see not only modern race cars being serviced but also classic racers and series-production models under restoration.

2017 911 RSR

During Porsche's media presentation at the 2016 Los Angeles Auto Show, it revealed its next-generation 911 GT

▲ The 2017 911 GTR, unveiled in Los Angeles, made its competition debut at the 24 Hours of Daytona. *Porsche Presse*

racer and its competition debut at the Daytona 24 Hours race in January 2017. Porsche Motorsport designers and engineers positioned the 4.0-liter, direct-fuel-injection, rigid-valve-drive engine ahead of the rear axle in true mid-engine configuration.

"The new 911 RSR is a completely new development," Motorsports director

▲ The 2018 911 GTS coupe reintroduced black satin trim from 1970s G-Series models. *Porsche Presse*

Dr. Frank-Steffen Walliser explained. "The suspension, body structure, aerodynamic concept, engine, and transmission have all been designed from scratch." The engine's amidships position allowed Porsche engineers to develop a large rear diffuser, something impossible with its previous rear-end location. "Combined with a top-mounted rear wing adapted from the LMP1 race car, the 919 Hybrid, the level of downforce and the aerodynamic efficiency were significantly improved."

Depending on the size of the restrictor, Walliser explained, the engine developed as much as 510 horsepower delivered to the ground through a sequential six-speed gearbox shifted with paddles at the steering wheel. The car competed in the FIA's LM GTE category during 2017 and 2018, with Porsche GT Team and Dempsey-Proton Racing claiming third place in the Pro and Am team Endurance Trophy standings, respectively.

2017–2018 911 GTS MODELS

Porsche carried on its typical progression and expansion of the 911/Typ 991 lineup, introducing the 911 GTS series in early January 2017; it was available in Germany in March and worldwide soon after. This series included a rear-wheel-drive 911 Carrera GTS coupe and cabriolet as well as a 911 Carrera 4 GTS coupe, cabriolet, and Targa. With a newly developed turbocharger, Porsche boasted engine output of 450 horsepower—30 more than the 911 Carrera S models offered.

The GTS lineup included Porsche Active Suspension Management (PASM) as standard equipment. Porsche quoted its best acceleration in the series with the rear-drive GTS, seven-speed PDK transmission, and Sport Chrono package, delivering 0 to 100 kilometers per hour (62.14 miles per hour) in 3.6 seconds.

YEAR	**2018-2020**
DESIGNATION	**Carrera GTS (rear-wheel drive) Carrera 4 GTS (all-wheel drive)**
SPECIFICATIONS	
MODEL AVAILABILITY	Carrera GTS Coupe, GTS Cabriolet Carrera 4 GTS Coupe, Carrera 4 Cabriolet Targa 4 GTS
WHEELBASE	2,450mm/96.5 inches
LENGTH	4,528 mm/178.3 inches
WIDTH	1,978 mm/77.9 inches
HEIGHT	1,297mm/51.0 in GTS Coupe 1,299mm/51.1 in 4 GTS Coupe 1,291mm/50.8 in GTS Cabriolet 1,293mm/50.9 in GTS 4 Cabriolet 1,291mm/50.8 in Targa 4 GTS
WEIGHT	1,450kg/3,197 lbs GTS Coupe (man) 1,470kg/3,241 lbs GTS Coupe (PDK) 1,495kg/3,296 lbs 4 GTS Coupe (man) 1,515kg/3,340 lbs 4 GTS Coupe (PDK) 1,520kg/3,352 lbs GTS Cabriolet (man) 1,540kg/3,396 lbs GTS Cabriolet (PDK) 1,565kg/3,451 lbs 4 GTS Cabriolet (man) 1,585kg/3,495 lbs 4 GTS Cabriolet (PDK) 1,585kg/3,495 lbs Targa 4 GTS (man) 1,605kg/3,539 lbs Targa 4 GTS (PDK))
TRACK FRONT	1,541mm/60.7 inches
TRACK REAR	1,544mm/60.8 inches
WHEELS FRONT	9J x 20 ET 51
WHEELS REAR	12J x 20 ET 63
TIRES FRONT	245/35 ZR 20
TIRES REAR	305/30 ZR 20
CONSTRUCTION	Monocoque Steel/Aluminum/Magnesium
SUSPENSION FRONT	MacPherson strut with anti-roll bar
SUSPENSION REAR	Aluminum multi-link with anti-roll bar
BRAKES	Discs 350x34mm front; 330x28mm rear
ENGINE TYPE	Horizontally opposed water-cooled DOHC six-cylinder
ENGINE DISPLACEMENT	2,965.78cc/180.98 CID
BORE AND STROKE	91x76mm/3.58x2.99 inches
HORSEPOWER	450@6,500 rpm 285@6100rpm (M64/21 1997-1998)
TORQUE	450@6,500 rpm
COMPRESSION	10.0: 1
FUEL DELIVERY	Direct fuel injection, twin turbochargers, intercoolers
TOP SPEED	(with 7-speed manual transmission; deduct 1 mph for PDK) GTS Coupe 193mph 4 GTS Coupe 192mph GTS Cabriolet 192mph 4 GTS Cabriolet 191mph Targa 4 GTS 191mph

While all the GTS models deliver a top speed in excess of 300 kilometers per hour (186.4 miles per hour), the status of fastest goes to the rear-drive GTS coupe with manual gearbox at an impressive 312 kilometers per hour (193.87 miles per hour).

Both the interior and exterior reminded longtime Porsche loyalists of the styling changes made starting with the G-series 1974 model year, when trim went from brightwork to satin black. This included the Targa bar, supplied in black for the first time since Typ 991 Targa introduction.

2018 911 GT3

First seen in March 2017, the next-generation 911 GT3 soon set a new lap record at the Nürburgring. On May 4, Porsche test driver Lars Kern, driving an early production example, turned a lap in 7 minutes 12.7 seconds over the Nordschleife. "A few years ago lap times like this could only ever be achieved by thoroughbred race cars with slick tires," GT Product Line director Andreas Preuninger said. "The new GT3 achieves this with comparatively modest power and is still fully suitable for everyday use."

▲ The redesigned top of the 2018 911 GTS Cabriolet faithfully follows the smooth arc of the coupe roofline. *Porsche Presse*

▲ Porsche's body designers resurrected not only satin black trim details but redid the Targa bar in the style of the 1974 G-series 911s. This is the 911 GTS4 Targa. *Porsche Presse*

▶ Test driver Lars Kern turned a 7:12.7 lap at the Nürburgring in this early-production 2018 911 GT3. *Porsche Presse*

ONE MILLION 911s

Porsche celebrated a significant accomplishment on May 11, 2017, when the one-millionth 911 left the Zuffenhausen assembly line. In 1950, when Ferry Porsche brought his company back to Zuffenhausen from Gmünd, Austria, he and his family wrestled with continuing any further auto manufacture. When he committed to acquiring five hundred car bodies for 356s from neighboring Reutter Karosserie, he worried whether the company would be able to sell those over a couple years.

By the time his designers Heinrich Klie and Gerhard Schröder completed the 911's designs under the supervision of Ferry's son F. A. "Butzi" Porsche, Ferry had to acquire Reutter, and—to keep paying a much-enlarged employee population—his workers assembled more than 3,300 of the cars in the first year of its life, 1965. In contrast, during model year 2017, Porsche assembled 47,467 of the 911s, more than fourteen times as many as in the car's inaugural year.

Porsche displayed the one-millionth 911 at the 24 Hours of Le Mans. During one of its "field trips" into the countryside, Porsche Presse took the car down the road to the village of Teloché, where Porsche headquartered its racing team in a small garage along Rue du 8 Mai. Porsche has used this facility since May 1951, when interim racing team manager Paul von Guilleaume hired the place for Porsche's needs.

2018 911 TURBO S EXCLUSIVE SERIES

Introduced June 7, 2017, this car came from the Porsche Exclusive Manufaktur in-house facility at Zuffenhausen. With 607 horsepower, the coupe accelerated from 0 to 100 kilometers (62.14 miles per hour) in 2.9 seconds, reached 200 kilometers (124.28 miles per hour) in 9.6,

▲ The one millionth 911 takes a drive down memory lane through the French village of Teloché where Porsche headquartered Le Mans efforts for decades. *Porsche Presse*

▼ The 2018 911 Turbo S Exclusive Series in Golden Yellow Metallic. *Porsche Press*

▲ The 2018 GT3 Touring has the rear wing replaced with an elevating spoiler with a "Gurney flap." *Porsche Presse*

▼ The 2018 911 Carrera T—less opulent, more sublime. *Porsche Presse*

and topped out at 330 kilometers per hour (205.05 miles per hour.)

Brake calipers for the PCCB system were available in black with the Porsche logo in Golden Yellow Metallic. Porsche Active Suspension Management (PASM), Sport Chrono, rear-axle steering, and Porsche Dynamic Chassis Control (PDCC) were standard equipment. The car introduced the Golden Yellow Metallic paint but was available in other colors as well. Porsche limited production to five hundred units.

2018 911 GT3 TOURING

On a calendar date particularly redolent for Porsche enthusiasts—September 11, 2017, or 9/11—Porsche introduced its newest 911 GT3 variation to media and guests at the Frankfurt Auto Show. "With the exception of the rear spoiler treatment, the body of the GT3 with the optional Touring Package remains unchanged," according to Porsche. The automatically extending rear spoiler from the 911 Carrera models was equipped with an aerodynamic tip ("Gurney flap") finished in body color.

The interior was leather instead of Alcantara, and seating surfaces were cloth. A six-speed manual gearbox is standard equipment. Porsche began deliveries to U.S. customers in early 2018 at a price of $143,000 excluding other options and delivery fees.

2018 911 CARRERA T

After resurrecting the "Touring" model designation from its Carrera RS heritage, one defining a more sublime version of the purely sporting 2.7-liter models, Porsche pulled another well-regarded letter from its past alphabetical history, the slightly less opulent 911 T, introducing it on October 23, 2017. But "less opulent" is less than accurate, as Porsche engineers and product planners

YEAR	**2018-2020**
DESIGNATION	**Carrera T (rear-wheel drive)**
SPECIFICATIONS	
MODEL AVAILABILITY	Coupe
WHEELBASE	2,450mm/96.5 inches
LENGTH	4,526mm/178.2 inches
WIDTH (w/mirrors)	1,978 mm/77.9 inches
HEIGHT	1,285mm/50.6 inches
WEIGHT	1,425kg/3,142 pounds manual 1,445kg/3,186 pounds PDK
TRACK FRONT	1,544mm/60.8 inches
TRACK REAR	1,519mm/59.8 inches
WHEELS FRONT	8.7Jx20 ET 49
WHEELS REAR	11.5Jx20 ET 76
TIRES FRONT	245/35 ZR 20
TIRES REAR	305/30 ZR 20
CONSTRUCTION	Monocoque Steel/Aluminum/Magnesium
SUSPENSION FRONT	MacPherson strut with steel springs and 10-millimeter lowering; Porsche Active Suspension Management (PASM) with electronically controlled vibration dampers; two manually selectable damping programs
SUSPENSION REAR	Lightweight aluminum multi-link with steel springs and 10-millimeter lowering; Porsche Active Suspension Management (PASM) with electronically controlled vibration dampers; two manually selectable damping programs
BRAKES	Discs 350x34mm front; 330x28mm rear
ENGINE TYPE	Horizontally opposed water-cooled DOHC six-cylinder
ENGINE DISPLACEMENT	2,981.39cc/181.94 CID
BORE AND STROKE	91x76.4mm/3.58x3.01 inches
HORSEPOWER	370@6,500 rpm
TORQUE	331lb-ft@1,700–5,000 rpm
COMPRESSION	10.0: 1
FUEL DELIVERY	Direct fuel injection, twin turbochargers, intercoolers
TOP SPEED	Manual 182mph PDK 180mph

▲ The interior of the 2018 911 Carrera T is well equipped, if simplified. *Porsche Presse*

endowed this model with features from the Carrera S models that had never been available on base 911 Carrera cars but that enthusiastic drivers desired. These included the standard 370-horsepower Carrera engine with PASM, Sport Chrono, a six-speed manual transmission with shortened shift lever, and optional rear-wheel steering. As with the GT3 and GT3 Touring, Porsche eliminated the rear seats as well as the Porsche Communication Management system (though both remain optional).

PORSCHE MUSEUM SPECIAL EXHIBITION

From December 14, 2017, through April 8, 2018, the Porsche Museum displayed one of its most recent and significant acquisitions: 1964 Typ 901 No. 057. Assembled in October 1964, the red coupe became a true barn-find hit when a German TV crew shooting an antiques and memorabilia program stumbled on it and another 911. The crew asked the museum for help identifying and evaluating the cars, and soon the museum staff acquired both, with price determined by an independent third-party expert.

For three years, museum mechanics painstakingly restored the very rough and rusty 901 to show condition, preserving as many original parts as possible and using "genuine body parts from the time taken from a different vehicle," Porsche Museum spokesperson Astrid Böttinger stated. "The general rule was to retain parts and fragments where possible rather than replacing them."

2018 PORSCHE TYP 911 GT3 RS

Porsche Motorsport unveiled the 2018 model 911 GT3 RS at the Geneva Motor Show on February 21, 2018. Its naturally aspirated 4.0-liter engine develops 520 horsepower, enough to accelerate the car from 0 to 100 kilometers per hour (62.14 miles per hour) in 3.2 seconds through its specially calibrated seven-speed PDK transmission; top speed is 312 kilometers per hour (194 miles per hour). Engineers modified the rear suspension, recalibrating the rear-axle steering system for maximum dynamics and precision.

Porsche also offered a Club Sport version of the car at no additional charge. This package included a rollover bar, a manual fire extinguisher, preparation for a battery-disconnect switch, and a six-point safety harness for the driver. One further option was available: the Weissach package, which incorporated additional carbon components for the chassis, interior, and exterior as well as optional magnesium wheels. The Weissach package delivered the lightest-weight variation of the GT3 RS, at 1,430 kilograms (3,153 pounds). It was available for orders at the Geneva Show with deliveries beginning in April 2018. Porsche quoted a price of €195,137, including VAT and country-specific equipment.

FIA WORLD ENDURANCE CHAMPIONSHIP SAFETY CARS

On April 4, 2018, Porsche announced its agreement with the FIA to provide a 2018 911 Turbo as race safety car for the Sports Car World Endurance Championship. The 3.8-liter automobile delivers 540 horsepower through all-wheel drive and remains mostly "stock" except for safety equipment that includes a rooftop light bar, a two-way radio for communications with race management, and brakes and suspension specially adjusted for racetrack needs. In addition, Porsche Motorsport fitted the car with full-shell racing seats with six-point harnesses for driver and passenger, a monitor that displays track signals and conditions, and flashing LED lights front and rear. With its seven-speed PDK, the Turbo accelerates from 0 to 100 kilometers per hour (62.14 miles per hour) in just 3 seconds and is capable of a top speed of 320 kilometers per hour (198.84 miles per hour). Porsche's loan of this vehicle continues through the 2020 racing season. One of the cars is set up in this way as "intervention car," with another fifteen vehicles from Porsche as part of its program with the FIA.

PORSCHE EXPERIENCE CENTER SHANGHAI

Porsche opened its sixth Experience Center in the world on April 27, 2018, right next door to the Shanghai International Circuit. The entire facility encompasses 100,000 square meters (1,076,390 square feet) and incorporates a handling course, off-road track, restau-

YEAR	2019
DESIGNATION	911.2

SPECIFICATIONS

MODEL AVAILABILITY	911 GT3, GT3 RS, GT3 Touring
WHEELBASE	2,453mm/96.57in
LENGTH	4,557mm/179.4in
WIDTH	1,978mm/77.87in (w/mirrors)
HEIGHT	1,297mm/51.06in
WEIGHT	1,430kg/3,153lb
TRACK FRONT	1,588mm/62.52in
TRACK REAR	1,557mm/61.30in
WHEELS FRONT	9.5Jx20RO50
WHEELS REAR	12.5Jx21RO48
TIRES FRONT	265/35ZR20
TIRES REAR	325/30ZR21
CONSTRUCTION	Monocoque Steel
SUSPENSION FRONT	McPherson spring strut axle with wheels suspended independently on trailing links and wishbones, split wishbones for camber adjustment, anti-roll bar, all suspension mounts ball jointed. Lightweight cylindrical coil springs with normal and sport mode, dampers with double clamping at wheel carrier
SUSPENSION REAR	Multi-link rear axle with wheels suspended independently on five control arms, split wishbones for camber adjustment, bodywork top mount as ball-joint mount, anti-roll bar, all suspension mounts ball-jointed. Barrel springs with integrated auxiliary spring with progressive characteristic. PASM adjustable damper system with normal and sport mode.
BRAKES	Dual circuit brake system, hydraulic with vacuum-controlled tandem brake booster. Standard system grey cast iron discs with aluminum brake rotor hubs front/rear. Six piston aluminum monobloc front calipers; four piston aluminum monobloc rear calipers; rotors internally ventilated and cross drilled front and rear. Optional brake system: Porsche Ceramic Composite Brakes (PCCB) with carbon fiber reinforced ceramic brake disc with aluminum rotor hubs front and rear; six piston aluminum monobloc fixed calipers in front, four piston fixed calipers in rear, yellow calipers, third generation ceramic brake discs.
ENGINE TYPE	Horizontally opposed water-cooled DOHC six-cylinder
ENGINE DISPLACEMENT	3,996cc/243.8CID
BORE AND STROKE	102.0x81.5mm/4.02x3.21in
HORSEPOWER	520@8,250rpm
TORQUE	346 lb-ft@6,000rpm
COMPRESSION	13.3:1
FUEL DELIVERY	Direct fuel injection
TOP SPEED	193 miles per hour

▲ The 2019 GT3 R tests at the Nürburgring Nordschleife. *Porsche Presse*

rant, café, meeting rooms, and Driver's Selection store. Throughout 2017, Porsche had delivered around 71,500 vehicles to Chinese buyers, exceeding the United States as its largest market. Of these around 13,900 were two-door sports cars, of which more than half were 911s.

2019 PORSCHE 911 GT3 R

Starting on May 11, 2018, Porsche accepted orders for its newest customer full-race 911, the rear-engine, rear-wheel-drive GT3 R. Its 4.0-liter engine provided racers 550 horsepower through extremely high-pressure direct fuel injection, variable intake- and exhaust-valve timing, and fuel feed through six throttle butterflies. Porsche's six-speed sequential constant-mesh transmission used an electronic shift actuator for fast, precise gear changes.

Cockpit redesign incorporated a fixed seat secured by six bolts with an adjustable foot-pedal box. This placed the driver as near as possible to the vehicle's center of gravity. Fully removable doors were reinforced with a side-impact element of carbon fiber, Kevlar, and aluminum construction with energy-absorbing plastic foam. For the first time the cockpit of the GT3 RS was air conditioned while still incorporating direct connections for driver's suit and helmet cooling as well.

The price was €459,000 plus country-specific VAT.

2019 911 SPEEDSTER CONCEPT

Porsche tantalized enthusiasts and customers on June 8, 2018, when it unveiled its latest Speedster concept in Zuffenhausen to kick off its 70 Years of Sports Cars celebrations. The car was full of new equipment, ideas, and innovations from Porsche Exclusive Manufaktur. Its naturally aspirated flat-six engine developed more than 500 horsepower and, in a nod to Speedster purity, was introduced with a six-speed manual transmission. Following display in Zuffenhausen, the concept traveled to the Goodwood Festival of Speed in the United Kingdom in July and on to Rennsport Reunion in California in September. Then, at the Paris Salon de l'Auto on October 1, Porsche announced plans to put the Speedster into production starting in 2019 and limit its run to 1,948 cars.

2019 TYP 935 CLUBSPORT

On September 27, 2018, during the Rennsport Reunion at Laguna Seca

▲ The 2019 911 Speedster saw limited production of 1,948 cars. *Porsche Presse*

Raceway in Monterey, California, Porsche announced a new Club Sport version Typ 935 with bodywork reminiscent of the legendary Typ 935/78 "Moby Dick." Porsche Motorsport planned to limit assembly to seventy-seven units of the 700-horsepower racer. "Because the car isn't homologated, engineers and designers didn't have to follow the usual rules and thus had freedom in the development," Motorsport director Dr. Frank-Steffen Walliser said at the unveiling. The 1,380-kilogram (3,042-pound) race car was available for immediate order for €701,948 plus country-specific VAT, and Porsche was to begin delivering the Clubsports at exclusive events starting in June 2019.

PORSCHE EXPERIENCE CENTER HOCKENHEIM

Porsche broke ground on its seventh Experience Center, with an opening planned for late 2019 inside the Hockenheim Grand Prix Circuit. It encompasses 160,000 square meters (1,722,225 square feet, approximately 39.6 acres), including tracks and structures similar in purpose to the PECs in Leipzig, Atlanta, Los Angeles, and elsewhere.

911 GT2 RS CLUBSPORT

Unveiled at the Los Angeles Auto Show on November 28, 2018, the GT2 Clubsport is intended for track days but also selected international motorsports events. This car shares the same 3.8-liter twin-turbocharged, 700-horsepower flat-six engine and rear-wheel-drive configuration as the limited-run Typ 935 introduced at Porsche's Rennsport Reunion. Porsche delivers this engine power to the ground via a rigidly mounted seven-speed PDK transmission spinning 310-millimeter-wide rear wheels. The Clubsport weighs 1,390 kilograms (3,064 pounds) and is equipped with Porsche Stability Management (PSM), traction control, and anti-lock braking. Porsche began accepting orders on the day of announcement, at €405,000 plus country-specific VAT, with delivery starting in May 2019.

▲ The 2019 Typ 935 Club Sport honors the 1978 935 racecar nicknamed "Moby Dick." *Porsche Presse*

▼ The 2019 911 GT2 RS Clubsport. *Porsche Presse*

2020 992 CARRERA 4 COUPE
2020 CABRIO
2020 CARRERA 4S
2020 CABRIO 4S
2021 911 TURBO AND TURBO S
2021 911 TARGA 4 AND TARGA 4S
2022 911 GT3 AND GT3 TOURING
2022 911 GTS
2023 911 T
2023 911 SPORT CLASSIC
2023 911 GT3 RS
2023 911 PORSCHE DESIGN TARGA 4S AND 911 GTS CABRIOLET AMERICA EDITION
2023 911 DAKAR
2024 911 S/T
2025 911 CARRERA AND 911 GTS (TYP 992.2)

CHAPTER 10
THE EIGHTH GENERATION 2020–ON

2020 911 TYP 992 CARRERA S AND 4S CABRIOLETS

Porsche introduced its eighth-generation 911, the Typ 992, to guests and media on the eve of the Los Angeles International Auto Show at its Porsche Experience Center Los Angeles on November 27, 2018. The 3.0-litered flat-six twin-turbocharged engine delivered 443 horsepower—an increase of 23 over the 991—as a result of improvements to the fuel-injection system and a better layout for turbos and intercoolers (moving them from behind the rear wheels toward center rear). Porsche projected the base Carrera engine to be turbocharged with output at 385 horsepower. Engineers also shifted the engine 20 millimeters (0.8 inches) forward on the platform to increase rigidity, and the 992 featured Porsche's new eight-speed PDK transmission. With this configuration, the rear-drive S accelerated from 0 to 60 miles per hour in 3.5 seconds and the all-wheel drive 4S in 3.4 seconds. Top track speed was reported as 191 miles per hour for the Carrera 4 and 190 for the 4S. Porsche introduced a seven-speed manual transmission in late 2020, bringing a savings of around 5 kilograms (55 pounds). The new PDK was not only heavier but also significantly larger due to empty space meant to house an electric motor should Porsche decide to offer a hybrid 911.

August Achleitner, retiring 911 development director, unveiled new driver's assistance technology called "Wet Mode," which sensed moisture on the pavement and tires and was standard on the new car, and an optional "Night Vision," which incorporated thermal imaging.

◀ Within months of introducing new Typ 992 coupes, Porsche unveiled corresponding Carrera and Carrera S Cabriolets in rear- and all-wheel-drive versions. *Porsche Presse*

▲ "The big challenge when you make a new 911 is to not make a new car," Porsche's chief of 911 development Michael Rösler said. "It always has to be recognizable as a 911." The 992 Carrera 4S managed that perfectly. *Porsche Presse*

As with other generations, wheel size increased, and the Typ 992s rode on 20-inch front wheels and 21-inch rears, a technology introduced on 991 GT3 RS and GT2 RS variants. The redesigned car body was 45 millimeters (1.77 inches) wider at the front than the Typ 991 to accommodate a 45-millimeter increase in front track, while rear body width increased to 1,852 millimeters (72.91 inches), which was the overall width of 991-generation Carrera 4 and GTS models. (This also enclosed a 44-millimeter increase in rear track width, and these changes made the track widths identical for both rear- and all-wheel-drive 911s). A significantly wider variable-position rear spoiler and seamless light bar highlighted the appearance of the back end of the new cars. Except for front and rear fascias, the car body was all aluminum. The 992 shared the wheelbase of the 991, at 2,450 millimeters (96.5 inches), and the basic platform was largely unchanged; overall length grew by barely 1 inch, with slightly longer overhangs. Achleitner estimated that 80 percent of the 992 was new and different from the 991, with most of the carryover in the engine.

Porsche design chief Michael Maurer explained the inspiration for the new body was modern reality—wider wheel track—and visual history, in particular the first generation 911 Turbos, for which Porsche didn't widen the car body to accommodate wider wheels and tires but simply extended the wheel wells and fenders. On the 992 it seemed to narrow the car's midsection. The front deck lid picked up noticeable design cues from the 911 G-series from the mid-1970s through the 1980s. Porsche's 911s from the 1970s also inspired the interior and the dashboard, anchored as always by the large, centrally mounted tachometer.

Porsche introduced the S at a base price of $113,200 and the 4S at $120,600, not including options, delivery fees, or applicable taxes.

Porsche planned to release its Typ 992 Cabriolet as well as rear-drive Carreras in 2019, with a new Targa following shortly after. And in typical succession, expected a Turbo, a Carrera T, and an assortment of GTS variations. Porsche Motorsport was developing GT3 models and, at this writing, will carry over the 4.0-liter normally aspirated flat six. Do not expect another 911 R, however.

By December 1, 2018, Porsche Cars North America had launched its "configurator" for the 992, though future variants were not included at launch.

▲ The 992's instrument panel and dashboard reminded many of early 911 interiors. *Porsche Presse*

With the PCNA configurator up and running by New Year's Day 2019, product planners quickly filled in new products. The Typ 992 911 Carrera S and 4S cabriolet arrived in January, and the base Carrera and Carrera 4 reached dealers' showrooms in coupe and cabriolet body styles the following July.

For buyers in the United States, the biggest news was the return of a manually shifted transmission with the eighth-generation Typ 992s. As mentioned before, Porsche's remarkable PDK gearbox, introduced with the 997.2 for 2009, brought on the steady decline in preference for self-shifting. By 2020, fewer than one in ten German buyers specified the manual. But in the United States, some 20 percent of the sales were manual gearboxes. *Car and Driver* magazine's K. C. Colwell explained the system in his June 2020 review: "The seven-speed stick is a transaxle that Porsche, with the help of trans-supplier ZF, converted from dual-clutch auto to manual in part with a crafty mechanism called MECOSA. That acronym stands for mechanically converted shift actuator, and

▼ Porsche Stability Management and Porsche Torque Management systems incorporated a new "Wet Mode" that activated a variety of acceleration, braking, and chassis alterations to manage risks of hydroplaning even with these fat 305/30 ZR21 tires fitted to the Carrera 4S. *Porsche Presse*

it makes the dual-clutch hardware with a conventional shift pattern."

The lightning-fast computer brains inside the PDK selected the optimum gear for the existing road speed, a situation that might not match the driver's intentions. Engine-protection software limited unloaded engine speed to 3,500 rpm, which, along with tires larger than the previous 997s, meant acceleration from 0 to 60 miles per hour took 3.6 seconds with the manual transmission as compared to 2.9 seconds for the PDK. But for those exclusively concerned with the driving experience, the active engagement of both hands *and* both feet was worth the extra 0.7 second getting on the freeway.

Porsche introduced the 2020 Typ 992 S starting at $114,650. Unlike the Tiptronic era, there was no difference between manual and PDK prices, but now the manual-equipped cars came with a full Sport Chrono package, mechanical limited-slip differential, and Porsche torque vectoring

▲ This Gentian Blue S cabriolet and all 911 Carrera S models ran with 443-horsepower opposed-six engines. The tiny PDK paddle sat forward on the center console. Cabriolet prices started at $127,450. *Porsche Presse*

◀ Porsche had Typ 992 Carrera S cabriolets, like this in Lava Orange, in dealerships in January 2020. U.S. customers were pleased to learn a revised seven-speed manual joined the eight-speed PDK as standard equipment. *Porsche Presse*

rear brake control. Porsche set the base price of the S cabriolet at $127,450.

The Carrera and Carrera cabriolet models offered buyers 379 horsepower with either seven-speed manual or eight-speed PDK, introduced at $98,750 for the coupe and $111,550 for the cabriolet. These 2021 models reached dealers in July 2020.

2021 911 TURBO AND TURBO S

Porsche introduced the Typ 992 Turbo and Turbo S in March 2020. These were vehicles that recalibrated the superlatives that typically started with "more" or "increased." As Porsche's media materials reported, "Currently, the most powerful 911 engine with two VTG [variable-turbine geometry] turbochargers delivers [640.8 horsepower].... With the Turbo-adapted eight-speed PDK, the sprint to [62.1 miles per hour] has been cut to 2.7 seconds (0.2 second faster than before) while the top speed remains [205.0 miles per hour]." This was a 67-horsepower increase. Engineers essentially reversed air routing to accomplish this gain. According to Porsche's media release: "Part of the process air and charge air cooling was swapped around: part of the process air now flows through the characteristic Turbo air intakes in the rear side sections. In front of the air filters, which are now situated in the rear wings, two other airflows through the rear lid grille have now been incorporated. This means the new Turbo S has four [air] intakes with a larger overall cross-section and lower wind resistance, improving engine efficiency."

For those ultra-serious about their performance driving, Porsche also offered a "PASM Sport Chassis" that engineers lowered 10 mm (0.39 inches). To assist with those high-lip driveways, an optional hydraulic system could lift the nose of the car 40mm (1.57 inches). Porsche's active rear-axle steering got a tune-up as well, improving the steering ratio by 6 percent to increase agility on those routes best known as "Porsche roads." Handling was further sharpened with Porsche adaptive aerodynamics, which introduced highly visible vertical-controlled cooling flaps at front and a larger but lighter rear wing that not only provided more rear downforce but also "acts as an air brake in the event of emergency braking at high speeds." Downforce peaked at 375 pounds. What's more, the front flaps served two functions: up to 43.5 miles per hour they opened to enhance cooling; from there up to 93 miles per hour, they closed to decrease drag and increase fuel consumption. Above 93 miles per hour, they reopened to improve aerodynamic balance.

▼ The 2021 Turbo S, here in Racing Yellow, has roadholding to handle 640 horsepower running uphill. But its standard-equipment ten-piston aluminum monobloc calipers and 420mm-diameter PCCB ceramic composite front brakes (four piston, 390mm rear) made the car just as fast downhill. In capable hands, the Turbo S Coupe delivered record-setting performance. On January 30, 2021, a similar coupe lapped Nürburgring's Nordschleife in 7 minutes 13.3 seconds using semi-slick tires, establishing a new best time for road-legal series-production automobiles. *Marc Urbano*

As has been the case since introduction of the Typ 964 Turbo in 1991, this new model once again grew in dimensions. These were not cosmetic changes to enhance the aggressive look of the Turbo but were necessary to house engine and chassis improvements. In capable hands, the car produced 1.10 G of lateral grip. For the 2025 model year, Porsche priced the Turbo S coupe at $230,400 and the cabriolet started at $245,195. The 572-horsepower 911 Turbo coupe listed at $199,195 and $211,995 for the cabriolet.

Car and Driver, ever America's journal giving voice to automotive fantasy, summarized the new Turbo S thusly: "There's a certain brutality as the blast of an engine tuned up to 171 horsepower per liter cranks through the drivetrain and into the road," Tony Quiroga wrote on their December 2020 website. "The tires screech. Passengers screech. Maybe even you screech—at least a little. Bear in mind that the tire squeal happens despite the Turbo S's standard all-wheel drive." (Porsche quoted Turbo S output at 640 horsepower at 6,750 rpm and 590 lb-ft of torque starting at 2,500 rpm.)

If America's Car and Driver magazine overstated, then Britain's CAR magazine notoriously understated and summed up the new Turbo S thusly: "Simply put, this model is more rounded than ever. At nine tenths, the driver can do almost no wrong, but the last step towards total satisfaction is a dare for man and machine, in exactly this order. No 911 has ever been this disarmingly multi-talented as this Turbo S," the magazine's "Car Test Team" wrote in the June 2002 issue.

▲ Turbos need not always be run at high speed. Sometimes the best use of a Turbo cabriolet is hurrying to the spot to watch a comet streak overhead or to pick out constellations. Marc Urbano

▼ Porsche introduced the 911 Turbo coupe at $199,195 and the cabriolet at $211,995. The Turbo S versions started at $230,400 and $245,195, respectively. As Britain's CAR magazine wrote, Porsche's Turbos provide "the sensation of bottomless speed and stability." Marc Urbano

"Those looking for a more focused driver are still better served elsewhere, but if the sensation of bottomless speed and stability appeals, look no further."

Porsche proved unequivocally the potential of this car on January 30, 2021. A car sponsored by German magazine *Sport Auto* lapped the Nürburgring Nordschleife in 7 minutes 17.3 seconds on semi-slick tires. This was the fastest road-legal series-production car lap to date.

Porsche further enhanced the myth and mystique of the Porsche in April 2022, introducing the Turbo S Lightweight. This $10,340 option included new lightweight, sound-insulated glass, one-piece carbon-fiber seats, new sport exhaust, rear seat delete, and the PASM lowering kit. With another option, the carbon-fiber roof, this saved 68 pounds over the normal Turbo S.

2021 911 TARGA 4 AND TARGA 4S

The Targa 4 and Targa 4S debuted on May 18, 2020. For those enamored of the Typ 991's innovative and entertaining cantilevering top system, fear not! Porsche retained the system, complete with its brushed silver or optional black Targa bar. Porsche also carried over its decision to assemble Targas strictly on the all-wheel-drive Carrera 4 and 4S platforms, providing, as they explained with the 991 series, all-weather safety and security to customers. These Porsche buyers were less concerned with track-day outings and preferred comfortable expeditions to five-star ski lodges in winter, beachfront resorts in summer, and fine dining anywhere in between. Targas shared technical details and accomplishments with the Carrera 4 and 4S coupes and cabriolets, including the choice of eight-speed PDK or seven-speed manual transmission at no additional charge. And extra-cost options were abundant.

A month later, the Targa 4S Heritage Design Edition broke cover, limited to 992 examples, and identified as the first of the 2021 model year Porsches. As Boris Apenbrink, director of Exclusiv Manufaktur vehicles at Porsche, explained, "Porsche set benchmarks in terms of design and styling from the outset—vehicles from different eras that are style icons nowadays." The Exclusiv Manufaktur division worked with the Style Porsche designer Grant Larson "reinterpreting exclusive 911 models with iconographic elements stemming from Porsche vehicles dating back to the 1950s through to the 1980s." Style Porsche's director of interior design Ivo van Hulten went further: "Individual memories that are brought back to life by certain color schemes, a feel, or patterns—this is a trend we are also witnessing in fashion or interior design and it forms the fundamental idea of our approach."

Exclusiv's choice of Cherry Metallic for an exterior color is evidence of this. Red has been in Porsche's color palette since the earliest days, and softly ribbed corduroy, the seating surface material of choice for the Heritage Targas, first

▼ Porsche's product planners envisioned their all-wheel drive Targas this way: top open, windows down, great comfort, and that rear spoiler up showing 911 owners that not all Dolomite Silver Targas—Dolomite Silver or otherwise—just cruise through scenery. *Porsche Presse*

▲ Bordeaux Red leather made a striking combination with Atacama Beige OLEA club leather for the Targa 4S Heritage Design Edition. Atacama Beige corduroy appeared on seat centers and door panels, tugging at heartstrings of older buyers who remembered Porsche's 1950s corduroy seats. *Marc Urbano*

appeared on seats in the 1952 Typ 356s. One other feature of the Heritage Targas tugged at longtime 911 owners' heartstrings: the classic five-spoke wheels Porsche introduced on its 1967 Typ 911 S were back. Originally cast in aluminum alloy by Otto Fuchs KG, each wheel saved 6.6 pounds over the steel wheels on standard 911s. But for the Heritage Targa, Exclusiv painted them high-gloss black. What's more, Porsche offered them across the entire 911 lineup as an option. "The 911 Targa 4S Heritage Design Edition is aimed at collectors and design-oriented enthusiasts alike with its unique combination of exquisite materials, detailed craftsmanship, and historical design elements from the 1950s and early 1960s," Apenbrink added.

2022 911 GT3 AND GT3 TOURING

Porsche tested the patience of their performance 911 loyalists. But everyone agreed the wait was worth it when the long-awaited seventh edition of the 911 GT3 and GT3 Touring models appeared in February 2021 as 2022 models. Deliveries began in May.

▼ Porsche's Exclusive Manufaktur division asked stylist Grant Larson to "reinterpret" the Targa, recalling signature design elements from earlier cars. Larson selected Cherry Metallic or an always-elegant Black for this Heritage Design, adding five-spoke wheels reminiscent of the earlier Fuchs models. *Marc Urbano*

"Developed in close collaboration with Porsche Motorsports, [the GT3] transfers pure racing technology into a production car more systematically than ever before," according to Porsche's media material for the car. "Its double-wishbone front suspension—the first time it has been used in a series-production Porsche—and sophisticated aerodynamics with

swan-neck rear wing mounting and striking diffuser originate in the successful 91 RSR GT racing car, and the [503 horsepower] 4.0-liter six-cylinder boxer engine is based on that of the 911 GT3 R, tried and tested in endurance racing. The dramatic-sounding, high-revving [9,000 rpm limited] engine is also used practically unchanged in the new 911 GT3 Cup."

That double-wishbone configuration was pure racing stuff. Adopted from open-wheel Formula One, Two, and Three racing cars, Porsche used it on its successful RS Spyder series beginning in 2005 and adapted it to the 911 for the RSR that won its class at Le Mans in 2017. Its claim to fame is legendary stability under lateral force assault. Couple this with the GT3's multi-link rear axle, rear steering, Porsche Stability Management (PSM), Porsche Active Suspension Management (PASM), lighter (but larger) forged alloy wheels, and tires that are 10mm (0.39 inch) wider on the tire patch than before, and the GT3 possesses "exemplary turn-in agility in combination with higher cornering performance and greater braking stability."

The new standard-equipment vented gray cast-iron brake rotors were dimpled now, instead of cross-drilled, the better to remove brake dust and strengthen braking action; they measured 16.06 inches compared to 14.96 inches on previous models. The optional Porsche Ceramic Composite Brake system utilized 16.14-inch front rotors and 15.35-inch rears.

All this blood pressure–enhancing text was fully supported with facts: a top speed of 198.8 miles per hour (or 197.6 miles per hour with PDK) and an acceleration to 62.0 miles per hour in 3.4 seconds with PDK. But don't despair over gearbox choice: "Porsche also offers the new model with a six-speed manual transmission for those who want a particularly purist driving experience."

▲ The double-wishbone front suspension was thoroughly proven in Porsche's 2005 RS Spyder racing car for its ability to manage roll and keep front tires vertical. Add in a multi-link rear axle, rear steering, PASM, and PSM and the result is superb handling on these 2021 Typ 911 GT3s. *Marc Urbano*

And for those who collect Nürburgring lap records, the 992 GT3 completed the 12.92-mile Nordschleife in 6:59.927, which was 17 seconds faster than its predecessor. All this arrived in a package rigorously weight managed to 3,126 pounds for the manual transmission model and 3,164 pounds with the PDK.

According to Porsche, the GT engines shared technology with their GT3 Cup racing siblings and those of the Porsche Mobil 1 Supercup. "On the intake side, for example, each of the six cylinders is provided with its own individual throttle valve at the end of the variable resonance intake system. This is positioned particularly close to the intake valves, thereby improving air supply and control precision."

The GT3's stunningly high engine speeds were possible due to "speed-resistant valve actuation [that] takes place via rigid rocker arms, which do without hydraulic clearance compensation. Porsche adjusted the correct valve clearance at the factory by means of interchangeable shims." There was no need for later adjustment.

The GT Cars team at Porsche wasn't concerned only with weight. Road-holding and handling have long been their primary target. "This starts with the optimized front end with its generously dimensioned openings," Porsche's media materials stated. "These route air in a targeted way to the front wheel brakes and radiators. This air is then discharged again behind the centrally positioned front radiator through a newly designed, now two-part air vent in the lightweight bonnet. The precise calculation of this air duct improves airflow and then increases cooling, while at the same time increasing aerodynamic downforce at the front axle.

"The front diffusers also benefit from this: in combination with the wide spoiler lip, they ensure a more constant flow of air along the fully clad vehicle underbody, particularly at

◀ As with all the 2022 GTS models, this GT Silver Targa mixed together elements from the GT3, Carrera S, and Turbo. GTS prices ranged from $138,050 for the coupe up to $158,150 for the Targa 4 variant. *Porsche Presse*

▼ Exclusiv Manufaktur dove deep into Porsche history to commemorate 50 years of the Targa. Exclusiv created an homage to the trendsetting design that Porsche Design's founder Butzi Porsche supervised, limited to 750 copies and Satin Platinum Black exterior paint. *Porsche Presse*

higher speeds. The new GT3 reaps the benefits of this aerodynamic design at the rear axle. Here, the fully functional rear diffuser supplemented by large fins accelerates the air directed at it in such a way that the vacuum generated sucks the high-performance sports auto onto the road even more powerfully. This has the significant advantage that downforce [is] produced in an especially efficient way because it hardly influences the car's drag." The downforce, including the effects of the swan-neck rear wing, "already exceeds the value of the previous model by 50 percent."

Porsche offered the car starting at $169,700 in the United States.

2022 911 GTS

Porsche followed the 911 T with the 992 version of the higher performance GTS models as coupe and cabriolets. The American magazine *Road & Track* described it as the 911 GT3 "for everyone else," and reviewer Brian Silvestro characterized it as "not quite a GT3, but not quite a Carrera 4S either. Simply put, it's the most capable version of the 'normal' 911." It was a hybrid, not in the sense that it had electric power

but that it blended bits and technologies from the Carrera S line, the GT3 group, and Porsche's Almighty Turbos, splashing in, or ladling on, upgrades and improvements where useful. The "S" power plant in the GTS delivered 473 horsepower and 420 lb-ft of torque. Planners pilfered the Turbo's center-lock wheels and their brakes as well as that extra 0.39-inch ride-height drop. The exterior trim was blackened; interior designers removed some sound insulation—but, in contrast to the GT3, they left some; they installed their heavily bolstered Sport Seat Plus, Porsche's most businesslike seat, before dropping into the no-compromise, slim-hips-only buckets; and actual push buttons activate exhaust flaps, nose lift, damper firmness, and climate control.

The GTS MacPherson front suspension configuration, coupled with the car's extremely responsive variable-ratio steering, "takes very little steering wheel angle to actually get the car turned. . . . Mix in the optional rear-axle steering on our tester, and it's easy to reach your own limits well before the car starts to approach its own," Silvestro noted.

Porsche configured the GTS on Carrera S and Carrera 4S platforms, that is, rear-wheel or all-wheel drive with seven-speed manual or eight-speed PDK transmissions. Product planners distributed the GTS wealth and goodwill broadly across the 911 lineup, offering the variation in coupe, cabriolet, and Targa 4 body styles. Porsche introduced the lineup as 2022 models, setting base prices as follows: Carrera GTS, $138,050; Carrera 4 GTS, $145,350; Carrera GTS cabriolet, $145,350; and Targa 4 GTS, $158,150.

Sometimes it's humbling to think about Porsche's longevity. Remember, in 1972, Ferry Porsche took outside advice and purged family from the company, hiring outside professionals to run his

▲ "In a world where every car has twice as much power as it knows what to do with, it's also refreshing to drive a car with tires and suspension that can easily handle it at full gallop," according to *Road & Track* magazine's Mack Hogan. *Marc Urbano*

company. Ferry's son F. A. best known as Butzi, was first out the door. He retreated to the family's homestead in Zell am See, Austria, where he set up a design consultancy known as Porsche Design. It started modestly, progressing from watches and sunglasses into luggage, garments, and sporting goods. By 2022, it was celebrating its own history with special- and limited-edition automobiles. For Butzi's shop, their choice was natural, the Targa, one of the 911 models over which Butzi supervised all elements of its appearance. In the late 2022 model year, Porsche released the Edition 50 Years Porsche Design Targa, limited to 750 examples and priced at $197,200. The Special Edition Targa was offered only in black, with satin platinum, Turbo S wheels, various body accents, and the Targa bar. To be completely faithful to its more humble beginnings, the company delivered each car with an updated Porsche Design Chronograph I, the timepiece Porsche claims was the world's first all-black watch.

2023 911 T

To spice up the middle of the 2023 model year, Porsche introduced . . . no, reintroduced a fewer-frills model from long ago, resurrecting the entry-level 911 T configuration. The designation first appeared in 1968, and Porsche resurrected it late in the 991 life cycle. Porsche's website model configurator was the car's best sales force: "The 911 Carrera T is a commitment to purism. A conscious release. For increased driving pleasure." It went on: "When contact with the road is even

▲ "Twisty canyon roads are where the car feels most at home," Drew Dorian wrote on *Car and Driver*'s website. The 2023 911 T, here in Gulf Blue, offered optional rear-axle steering for even more fun in the mountains. *Marc Urbano*

▼ Seven speeds, three pedals. Many enthusiasts consider Porsche's 911 base Carrera its best—purest—driving experience. Engineers studied that car and then made this one. Sport Chrono was standard, as was a mechanical limited slip; PDK is a no-cost option. *Marc Urbano*

more important. When every pound less means more agility. That's when sportsmanship takes over." Deleting rear seats, reducing insulation, and lightening side glass and the batter saved 77.2 pounds over the base 911 Carrera, bringing the T in at 3,241 pounds.

Porsche engineers adopted some of the chassis options they made standard on Turbo and GT3 models, dropping the already 0.39-inch-lowered PASM Sport Suspension another 0.39 inch. "The springs are harder and shorter, the anti-roll bars on the front and rear axles are stiffer. The spring rates have been significantly raised, making the 911 Carrera T even sportier on the road." Perhaps Porsche meant that designation "911 T"—which stands for "Touring"— to represent "the True 911." The T came equipped with either the seven-speed manual transmission with a shortened gear shift lever or the quicker-shifting eight-speed PDK. And just so the driver and his or her one passenger (rear seats are deleted) didn't mistake the T for any other car, the sport exhaust system "brings impressive resonance and an intense sports car sound" to the car.

Porsche's standard 2,981cc dual-overhead-camshaft opposed six-cylinder with direct fuel injection, twin turbos, and intercoolers, delivered 379 horsepower at 6,500 rpm. Acceleration to 60 miles per hour took 4.3 seconds with the manual or 3.8 seconds with the PDK, both with the standard-equipment Sport Chrono packages and a mechanical limited-slip differential. Porsche quoted top speed at 181 miles per hour.

One of *Road & Track* magazine's reviewers summarized the common-sense appeal of the 911 T. Mack Hogan first drove the car as part of a fleet test for their car of the year title. He noted that its 379 horsepower was far from the most powerful on offer. However, "what it does provide is sports car speed. For 98 percent of drivers, that's the better pace. It's fast enough to push you back in your seat, but slow enough that your mind can keep up with it." He continued, "In a world where every car has twice as much power as it knows what to do with, it's also refreshing to

▲ This Carmine Red 2023 911 GTS is "a GT3 for everyone else," *Road & Track* writer Brian Silvestro suggested. "Not quite a GT3, but not quite a Carrera 4S either. Simply put, it's the most capable version of the 'normal' 911." *Porsche Presse*

▼ The variable-ratio steering and turn-in was quick. "Mix in the optional rear-axle steering on our [Carmine Red 2023 GTS cabriolet] and it's easy to reach your own limits well before the car starts to approach its own," Brian Silvestro noted. *Porsche Presse*

drive a car with tires and suspension that can easily handle it at full gallop."

R&T's sometimes rival *Car and Driver* agreed: "Twisty canyon roads are where the car feels most at home," according to Drew Dorian on their October 19, 2023, website review. "Steering is crisp and communicative, and so is the manual transmission, which has an unambiguous clutch take-up point. Although these controls feel entirely natural, the brakes take some getting used to. Early in the pedal travel, the stoppers bite aggressively and smooth modulation is a learned behavior."

Porsche introduced the 911 T at $118,050 base price and offered an option for front lift system, rear-axle steering, and an extended-range 23.7-gallon fuel tank, among other items.

2023 911 SPORT CLASSIC

In April, Porsche unveiled its 2023 Typ 911 Sport Classic, a limited-production coupe—250 units—with innumerable special features redolent of the early 911 S models. With Porsche's new ability to install carbon-fiber roofs, this ultra-stylish model introduced the "double bubble" with subtle swells over the driver's and passenger's heads and a shallow valley between them. In Porsche's web materials, it summed up the new car classically: "Back to the Future." Visually inspired by the 1973 911 Carrera RS 2.7, the Sport Classic concept was first introduced as the 2009 Typ 997 Sport Classic (not available to U.S. customers). This latest version, based on the 992 platform, began deliveries in Europe in July 2022 starting in Germany and spread throughout the rest of the world starting in September. Porsche priced the car from $274,750. It followed the 2020 911 Targa 4S Heritage Design Edition as the second complete vehicle from Porsche Exclusiv Manufaktur and, according to Alexander Fabig, head of Customization and Classic, it blended the 911 Turbo 74.8-inch-wide body (though without the rear wing air intakes) with "a fixed rear spoiler in the style of the legendary ducktail of the Carrera RS 2.7, and the double-bubble roof."

Exclusiv Manufaktur manager Boris Apenbrink and Style Porsche designer Grant Larson offered few colors for the new Sport Classic: Sport Grey Metallic was the primary color, inspired by an early 356 color called Fashion Grey; Agate Grey Metallic, Gentian Blue Metallic, and solid black followed. The interior's Pepita houndstooth-style fabric covered seat inserts and door panels, reminiscent of the earliest 911s and 912s from 1965 and on.

There was nothing retro about the rear-drive drivetrain and its output. The 3,743cc twin-turbo flat-six produced 543 horsepower and 442 lb-ft torque and, through its seven-speed manual transmission, took the Heritage Targa to a top speed of 196 miles per hour. To aid classic drivers who lost their heel-and-toe driving technique to effortless PDKs, engineers fitted the gearbox with Porsche's auto-blip function, now optional on many models, that automatically matched engine revs to road speeds on downshifts. It was intensely amusing and greatly beneficial. Then, to keep handling where its sophisticated, experienced buyers expected it to be, engineers mated the standard PASM as used in the 91 Turbo and GTS range together with rear steering and sports suspension, which provided the 0.39-inch lower ride height.

▼ With "Classic" inspiring this retro 911's look, the Sport Grey Metallic paint evolved from a popular 356 color in the 1950s called Fashion Grey. New manufacturing capabilities led Zuffenhausen to install carbon-fiber roofs, permitting this striking "double bubble" configuration. *Porsche Presse*

▲ "The Sport Classic is more about winding roads, driver involvement, and a rollicking good time," according to Car and Driver magazine's November 2023 webpage. The interior evoked the mid-1960s with an update of a classic Pepita houndstooth cloth. *Porsche Presse*

On their November 2023 webpage, Car and Driver magazine's technical editor reminded readers, "A rear-drive manual was never going to rule the drag strip. The Sport Classic is more about winding roads, driver involvement, and a rollicking good time." Cornering delighted him: "Turning in, the Sport Classic is tenacious. Its front tires deliver a surprising surplus of grip that can tighten the arc even when you think you've overcooked it," he wrote. "As you roll onto the throttle at exit and revs build, a satisfying wiggle makes clear this is a rear-drive machine, but the ass end never hints at walking you tail first toward the guardrail. . . . The feel and feedback of it are all but impeccable."

2023 911 GT3 RS

One month later, in August 2022, just as the 2023 model year began, Porsche brought out its 991 GT3 RS with its expected wings and exciting graphics, a more visibly bold car than the standard GT3 and much more so than the almost stealthy 911 GT3 Touring. Prices for the RS started at $241,300 and for the Touring edition at $171,150. The two vehicles couldn't possibly represent more different answers to the definition of Grand Touring.

Eric Tingwell, writing for *MotorTrend* on their December 2023 website, noted, "In a world where automakers can't stop blathering on about no-compromise vehicles, the newest GT3 RS is gloriously, unapologetically committed to doing a single thing—cracking off incomprehensibly fast lap times—at the expense of all others." Speaking of which, on October 13, 2022, Porsche team driver Jörg Bergmeister shot around the Nordschleife in a GT3 RS in 6 minutes 49.328 seconds, setting yet another new production car record.

With 518 horsepower and 342 lb-ft of torque, the GT3 RS "can be so neutral with such extreme limits yet still feel approachable and unintimidating. Give credit to the perfectly calibrated controls for that. The steering seemingly runs on the same 70 millivolts as your nervous system, bypassing conscious thought as it turns feedback from the front wheels into subtle rolls and flicks of your wrists. The brake pedal, too, offers textbook-perfect feel and travel that makes it easy to feather your foot against the ABS threshold." With all the discipline Porsche GT Cars manager Andreas Preuninger and his team applied to this car, it weighed 30 pounds less than a similarly outfitted GT3. But at 125 miles per hour on a track, the car was 900 pounds heavier due to downforce. At

▼ Wings, flaps, brutal body cuts, and perforations, all in the pursuit of airflow management to great maximum downforce while controlling drag. If the GT3 Touring is sublime, this Weissach Package is strictly purposeful. *Marc Urbano*

◀ Despite an interior that looked seriously comfortable, Eric Tingwell from *MotorTrend* magazine explained, "The newest GT3 RS is gloriously, unapologetically committed to doing a single thing—cracking off incomprehensibly fast lap times—at the expense of all others." *Marc Urbano*

▼ Airflow management required time in sophisticated computer modeling programs and time with a smoke wand in a wind tunnel. Complex flow patterns kept the 2023 911 GT3 RS on the ground. Some airflow passed through the body twice. *Porsche Presse*

terminal velocity of 177 miles per hour with the aerodynamic settings in "high downforce" mode, the wings and flaps added 1,900 pounds to the weight of the car standing still.

Speaking of standing still, that same extremely sophisticated active aerodynamic system (along with the fat standard Michelin Pilot Sport Cup 2R tires and vented, cross-drilled ceramic disc brakes) brought the car from American freeway 70 miles per hour speeds to a stop in 133 feet. The rear wing, more than 4 feet in the air, pivoted to vertical as a highly effective air brake.

As *Car and Driver*'s K. C. Colwell put it in his review on the August 2023 website, "The 911 GT3 RS is a surprisingly civilized race car for the road."

In contrast, an enthusiast might also say that the GT3 Touring Package is a surprisingly capable road car for the track, though that is the opposite of its intent. Porsche on its website described the car like this: "The innate preserve of the 911 GT3 with Touring Package is the open road—ideally a terrain full of challenging bends. . . . It will not be instantly recognizable to everyone on the road. But that is precisely what increases its puristic appeal. And the experience of driving it."

Porsche gave a pair of GT3s, one with wing and the other without, to a pair of writers for *Panorama* magazine, the Porsche Club of America's monthly print journal. And, perhaps better to confuse than clarify issues, the winged car had the six-speed manual while the Touring had the PDK transmission. One of the *Panorama* testers was 35-year-old digital media coordinator Damon Lowney, and the other was 56-year-old technical director Manny Alban, a pair literally a generation apart. They had just a single day and no chance of time on a racetrack.

In a supreme testament to the GT Cars department and its hard work, Alban experienced a revelatory moment: "At one point when filming our video as we were test driving I forgot I was in the Touring and discovered that only when I glanced in the rearview mirror . . . it certainly reinforced my thoughts that the cars truly were identical." Prior to driving the cars, Alban believed he'd chose the winged GT3 if given the choice. After the drive, he wondered whether he had "matured as I've gotten older. I didn't need that boy-racer Cup car look, but instead preferred the traditional lines of the 911. Had we driven it on the track and I knew that a big chunk of my time as an owner of this car would be on track, then the GT3 with the wing would probably have been the winner."

The transmission choices vexed Lowney through the day. "We found it somewhat ironic that the GT3 had the manual while the Touring had the automatic," he wrote in the August 23 print edition. "Recall 2013 when Porsche announced the 991.1-generation would be PDK-only because it was faster on track. And then 2017, when the automaker announced the new 991.2 GT3: Customers could opt for a six-speed manual or seven-speed PDK for the first time, but the GT3 Touring would be available only in manual. Porsche's reasoning was that those who opted for the Touring Package wanted maximum engagement with the car, even at the expense of a few tenths of a second on a track." Lowney's observation: "Having a choice is the best choice." Then he veered toward his verdict: "First, a standard GT3 is best served with the six-speed manual. Second, a GT3 Touring is better with PDK. And third, well, I'm not going to argue with Porsche's decision to make the GT3 RS solely PDK. It's the only one of these three cars whose purpose is to shave lap times."

For another perspective, look no further than the January 2024 *Road & Track* magazine webpage on which Matt Farah offered this: "If you don't know cars, it's just another 911," he wrote. "If you do know cars, or even better, if you've spent time wheeling one, you understand why people are willing to pay tens, if not hundreds, of thousands of dollars over MSRP to get a GT3 Touring. Ferrari, Lamborghini, Audi, and Aston Martin have all but abandoned manual transmissions and natural aspiration, but not Porsche. If you want a brand-new exotic in which to row your own, Porsche is the last, best game in town. The irony is, that for all its talents, the GT3 Touring is not particularly adept at . . . touring."

And that is due to the exceptional job Preuninger et al. have accomplished in reproducing all the winged GT3 traits, characteristics, habits, pluses, and minuses in the flat-back version. And that, honestly, is part of what makes Porsche such a smart and successful company. With the addition of the GT3 Touring, the 992 lineup offered countless variations of the rear-wheel drive, all-wheel drive, manual gearbox/PDK transmission, open car/closed coupe (and even Targa). If a customer desires the ultimate driving experience, there is the GT3 or even GT3 RS; if another customer wants uncompromised driving interaction with the road but wishes to stay below (literally) the (police) radar, there's the GT3 with Touring Package. Porsche priced the car starting at $171,150.

2023 911 PORSCHE DESIGN TARGA 4S AND 911 GTS CABRIOLET AMERICA EDITION

A couple more specials appeared in time to cross over into the new year with the late 2022 introduction as a 2023

▼ Porsche's commitment to GT3 RS improvements led it back to the Nürburgring Nordschleife. On October 13, 2022, factory driver Jörg Bergmeister fired up this 911 GT3 RS and got around the Nordschleife in 6 minutes 49.328 seconds, establishing a new production car record. *Porsche Presse*

model. After the company offered the 750-example Edition 50 Years Porsche Design Targa, an even more exclusive 911 GTS Cabriolet America Edition appeared, limited to just 115 examples. These and the Sport Classic became instant collectibles for those fortunate enough to have ordered them.

In a very specific acknowledgment of American driver preferences, Porsche offered this latest commemorative as a 543-horsepower rear-wheel-drive model fitted with the seven-speed manual transmission. It marked the third time Porsche had specifically targeted U.S. buyers with a model tailored to tastes and demands of the time. The first was the racing-inspired 1952 America Roadster, based on a Typ 356 platform so significantly modified it earned its own Typ number, 540. California distributor Johnny von Neumann had begged U.S. importer Max Hoffman for a car he—and his fellow Californians—could race. This 20-car series was the progenitor of the Speedster, which arrived as a 1955 model and was the direct result of more begging from von Neumann. But this roadster soon became the must-have car of a generation of young Hollywood actors, giving Porsche an unanticipated sales boost among mid-1950s opinion makers.

Porsche had resurrected the America Roadster in 1992 on the Typ 964 platform. Total production reached 702 units because Porsche also sold the car throughout the rest of the world as the Carrera 2 Cabriolet Turbolook, while in the United States, it was a wide-body roadster strictly intended for cruising. Just 250 copies made it to the United States.

The new 2023 America Cabriolet appeared during the June 2022 Porsche Parade in Poconos, Pennsylvania. Porsche developed the car on the rear-wheel-drive GTS platform with its 473-horsepower opposed six-cylinder engine. Planners made PASM Sport standard, with its 10mm (0.39-inch) lower ride height. Buyers could option rear-axle steering and the PCCB ceramic brakes. As with the Porsche Design Targa, designers limited the America to a single color named Azure Blue 356. Porsche limited production to 115 cars—100 to the United States and 15 for Canadian purchasers—and deliveries in the United States began in late 2022, priced at $186,370.

2023 911 DAKAR

By this time, a few niche outsiders had created a market for off-road-style 911s. These were primarily preowned Typ 964 Carrera 4s converted to more rugged use. For model year 2023, Porsche introduced its own rough-and-tumble take-it-off-road-and-get-it really-dirty 911 as the 911 Dakar!

There were a number of features that made this 911 dramatically different from everything else in the 992.2 lineup, not least of which was the fact that many photos on Porsche's website or advertising showed the car dirty. Then there's

▼ Inspired by its own history and motivated by outsiders creating similar off-roading 911s, Porsche took its 911 GTS and made it into the 911 Dakar, a car with dramatically increased ride height and the manufacturer's blessing to get it dirty. *Porsche Presse*

◀ Porsche offered the Dakar in several racing liveries, including this color scheme reminding enthusiasts of the 1984 Typ 953 that racer Jacky Ickx entered in Porsche's first try at the Paris–Dakar desert race. One of Ickx's teammates was first overall. *Porsche Presse*

▼ The 911 Dakar interior was much more comfortable than the 1984 race car, complete with air conditioning, sound and navigation systems, and Porsche's PDK transmission. Its GTS power train gave it 473 horsepower, and rear-axle steering was standard equipment. *Porsche Presse*

ride height. Nearly every 992 model was available with PASM, which allowed owners to reduce ride height by 10-mm (0.39 inch) for improved handling and roadholding, especially the Typ 992 PDK-equipped Carrera 4 GTS, on which this car was solidly based. Some models described above went further with PSM Sport that lowered the car another 10 mm (0.39 inch). To quote Porsche's configurator: "Keep your distance. Not just at the front and rear—but also underneath. The Dakar is the only 911 that has a sport chassis to make it [1.97 inches] higher than the standard model. If space still gets tight under the stainless steel guard while on gravel roads, the 911 Dakar's specially designed lift system can raise the car by a further [1.18 inches]. The maximum speed at [3.15 inches] is [106 miles per hour]."

Other revisions seemed—on first examination—intended simply to make drivers giggle uncontrollably: "The Sport Chrono package for the 911 Dakar includes two further specific performance setups. Rallye mode—for gravel roads, muddy tracks, and wet grass—and Off-Road (with traction set at 50:50 front/rear) for tackling sand dunes and rocky passages. The Rallye Launch Control is also on hand to ensure a perfect rally stage start on loose ground"—in other words, to produce four perfectly matched rooster-tail trails of mud, dirt, sand, or other loose material.

At this point, it is well worth remembering that, since before the birth of the Typ 901/911 in 1964/1965, Porsche has encouraged customer rally participation. In those earliest days in Europe, international events crossed several national borders and spent much of their time on unpaved roads. By the early 1970s, customer racing offered welded-in roll bars, on-board fire extinguishers, underbody trays and impact shields, auxiliary lighting, optional tires, and replacement pieces and panels in other stronger and/or lighter materials. Porsche carried this on with the Dakar, offering a matte-black powder-coated roof rack with integrated LED auxiliary lamps, a 10-liter fuel and 12-liter water canister (designed to fit into the roof basket), a folding shovel incorporating an ax and a saw, "recovery boards" to place under the wheels and tires in soft sand or mud to aid in escape (also roof mounted), and other items.

Porsche offered the Dakar in five solid colors, plus a two-tone white over Gentian Blue Metallic (a paint scheme reminiscent of the legendary 1984–1986 Paris–Dakar Raid entries), as well as any color imaginable by Paint to Sample. Standard wheels were Satin Black–painted five-spoke Zoll Dakars with 19s in front and 20s in the rear. Rear steering was standard.

Porsche fitted the car with its 2,981cc twin-turbo flat-six that developed 473 horsepower at 6,500 rpm and 420 lb-ft torque between 2,300 and 5,100 rpm. Porsche quoted a top speed for 149 miles per hour and acceleration from standstill to 62 miles per hour in 3.4 seconds. Engineers fitted its engine in GT3 motor mounts. Limited to 2,500 units, Dakar prices started at $223,450 and soared from there with options and dealer markups contributing to price inflation. The run of cars sold out, and by August 2024, cars with between 10 and 1,000 miles on the odometer were available for $350,000 to $365,000.

What about real-world reactions to the car? "I just spent the day tearing down desert two-tracks and surfing dunes . . . in a factory Porsche 911. It's so preposterous, all we could do was laugh like a kid the entire way," *MotorTrend* magazine's Scott Evans wrote in his March 2024 website review. Or, "the Porsche 911 Dakar has rearranged our understanding of what a 911 can and should be cable of," according to *CAR* magazine's Ben Miller in his February 2024 website review following his introductory drive in Morocco. "We glide between tufts of elephant grass and skeletal long-dead trees, filling the air with dust, sand, and, where the car's underbelly (armored to an extent with CFRP-reinforced panels) clonks off the landscape, with expletives. What's more, we don't get stuck. And even as I sweat, I'm smiling."

After reminding readers, "Remarkably, the 911 Dakar is 911 GT3 RS money," and pointing out the obvious intended-destinations difference between the two, Miller recommended taking both these cars at face value, driving each "as intended. Do that and you're left open-mouthed at the bandwidth of Porsche's rear-engined icon and just how brilliantly its idiosyncratic dynamics work in environments as diverse as Silverstone and Sahara."

2024 911 S/T

Porsche's product planners reintroduced another legendary—and near mythical—model, the 911 S/T. Porsche revealed the car in August 2023 as a 2024 model and limited production to 1,963 cars, which was perhaps 30 times the number Porsche's racing shops had assembled for known racers between 1970 and 1973. Deliveries began just in time for Christmas, with prices starting at $290,000.

In some ways, the car might be considered as the GT3 RS Touring. The GT cars engineers at Flacht incorporated the RS's 3,996cc, 518-horsepower, naturally aspirated engine, the GT3's short-ratio six-speed GT manual gearbox, lightweight clutch, single-mass flywheel, extensive use of carbon fiber–reinforced plastic body panels, CFRP bucket seats, lightweight glass, magnesium wheels, insulation reduced to nearly nothing, and other severe disciplines that trimmed the S/T's weight to 3,056 pounds, the lightest of any 992. It adopted the GT3 RS double-wishbone front suspension and incorporated a CFRP lateral stabilizer to restrain wheel tilt angle under extreme cornering loads, ensuring the maximum tire patch for precise steering and unmatched braking. As Porsche's website explained, "Each 911 demands its own tuning. With the 911 GT3 RS, the focus is on the circuit. The 911 Turbo combines comfort and sportiness, and with the 911 S/T, pure driving pleasure." Porsche published a top speed of 186 miles per hour and acceleration from standstill to 60 miles per hour in 3.5

▼ Porsche's 2024 911 S/T was another model inspired by racing successes. The original fitted powerful 911 S drivetrains beneath lighter weight 911 T body shells, hence the name. The new car was more like a 911 GT3 RS Touring Package. *Porsche Presse*

seconds. (The S/T ground clearance was set at 3.9 inches, in stark contrast to its equally uncompromised cousin, at which height the 911 Dakar began its ascent.)

The car did not offer rear-axle steering, a decision that required a year of chassis and steering development to reconcile to the GT3 geometry (which, according to Preuninger, was conceived to work with rear-axle steering). This

▲ Revisions to the GT3 RS "allows the [S/T's] suspension to articulate more, especially in the rear. This car will move around on you," according to *Road & Track* writer Brian Silvestro. "You feel every minute change in direction." *Porsche Presse*

◀ Porsche described the S/T as "a strictly limited anniversary model—60 Years of the 911." Typical of anniversary commemoratives, every detail was meticulously considered for quality, weight, durability, and eye appeal inside and outside. *Porsche Presse*

▼ "We still have the idea and the demand on the car to feel super light and to shrink about you when you drive it, and that was the main goal" of the 911 S/T as Porsche's GT Cars godfather, Andreas Preuninger explained. *Porsche Presse*

▲ Surprising in so many ways! The 2025 911 GTS was the first Typ 992.2, and typical of past facelifts, the exterior differences were subtle; everything startling was beneath the skin on this Carmine Red coupe. *Porsche Presse*

▼ Everyone talked about the GTS's front flaps, seen here on this Ice Grey Metallic cabriolet. They opened to improve engine and brake cooling or closed to reduce midrange drag for better aerodynamics. They function automatically but with manual override. *Porsche Presse*

produced an interesting result, according to *Road & Track* writer Brian Silvestro on the magazine's September 2023 website: "Rather than being stiffer or more on-edge, the S/T pivots to a comparatively relaxed chassis setup that allows the car's suspension to articulate more, especially in the rear. This car will move around on you, and slices through corner in a less frantic way versus its GT-badged siblings. You feel every minute change in direction and shift in weight, yet it never feels uncomposed, no matter what sort of turn you [chuck] at it. It's made for real roads, with bumps, imperfections, and potholes. Not race tracks."

The hood, fenders, doors, and double-bubble roof—each in CFRP—were all painted in the exterior body color, which they limited to either black or Shore Blue Metallic (which was only offered with the S/T's optional Heritage Design package). Porsche characterized the S/T as "a strictly limited anniversary model—60 Years of the 911," and as such, its Exclusiv Manufaktur divison's Heritage Design Package incorporated numerous understated—but rigorously weight-

◀ Porsche's new 9A3B6 3.6-liter opposed six used a single turbo (extreme lower right corner) and a complex electric motor (just forward of the rear wheels) with a rotating electric motor that resembled a massive flywheel developing 54 horsepower. *Porsche Presse*

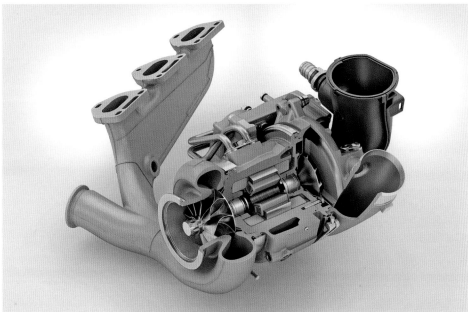

▼ The Turbo did double duty. A built-in electric motor kept turbines spinning when throttles closed, reducing full-throttle spool-up from three seconds to nearly zero. It eliminated the waste gate, siphoning off excess turbine speed into electricity generation. *Porsche Presse*

managed—interior and exterior design and styling details, ranging from two-tone-black-and-Classic Cognac interior appointments to incorporating original Porsche crests from 1963 to gold-colored model designation plates inside and out.

Andreas Preuninger's engineers incorporated an automatically extending rear spoiler to meet downforce demands with minimal air drag. The trailing edge of the spoiler utilized one of motorsports' subtlest innovations, the "Gurney Flap," a modest rigid CFRP lip named for its inventor, former 1960 Porsche Formula One racer Dan Gurney, that produced rear-axle downforce out of all proportion with its appearance. Beneath the rear of the car, a large-vane diffuser created a vacuum, increasing rear grip and sucking the S/T tightly to the road.

"What makes the 911 S/T so great can be understood by examining what it doesn't have, rather than what it does," *Road & Track* magazine reviewer Brian Silvestro concluded on the magazine's September 2023 website. "It takes just one look at the steering wheel to get it. There are no rotary knobs for drive modes or suspension settings. There are no paddle shifters for a dual-clutch transmission. Look at the body and you won't find a gigantic rear wing or ankle-slicing canards up front. Peek underneath, and things like AWD or rear-steer are nowhere to be found. The 911 S/T is the modern 911 experience in its purest, most exciting form." It's not a track car but one for real roads.

But the final verdict must go to the GT Cars director, Andreas Preuninger: "The 992 has gotten [to be] a pretty big car, everybody knows that. But we still have the idea and the demand on the car to feel super light and to shrink about you when you drive it and that was the main goal." Porsche introduced the S/T at $291,650 with production limited to 1,963 cars as a 2024 model.

2025 911 CARRERA AND 911 GTS (TYP 992.2)

Porsche has never shied away from launching bombshells into public expectations of the company. This started with the 356 (What? It's rear engined and air-cooled . . . ?) and continued with the 911 (Wait! It's still rear engined and air-cooled?). The real gobsmacker was the Cayenne sport utility vehicle introduced in model year 2000. That it sold like fresh water in a desert surprised only

◀ The 992.2 GTS interior looked familiar, but subtle changes were everywhere. Ever-shrinking steering wheels forced designers and engineers to confront difficulties in viewing the traditional five-gauge instrument panel, which was reconfigured. *Porsche Presse*

people outside Porsche management. Its success encouraged other surprises, a sedan, a second SUV, and a second sedan strictly electrically powered. Even as it was converting its two SUVs to all-electric drive in the coming years, in late May 2024 Porsche introduced its first series-production electric sports car, the 2025 911 GTS T-Hybrid. Electric sports car is alarmist, and inaccurate. It's a regenerative hybrid that produced 532 horsepower. According to Michael Rösler, chief of 911 development, "The big challenge when you make a new 911 is to not make a new *car*. It always has to be recognizable as a 911." At least, on its exterior. What was inside the skin made the new 911 perhaps *the* conversation starter during summer 2024, a time that otherwise might have concentrated on innovations the facelifted Typ 992.2 series offered. What's more, it upended the familiar order of things: the GTS normally arrived at the end of the 911 cycle, yet here it was, launching the 992.2 series!

Porsche press materials make it clear that—at least for now—the 911 GTS line was the only Hybrid in the 911 family. As Porsche introduced the Hybrid, it also presented the Typ 992.2 Carrera coupe and cabriolet. These are strictly internal combustion engine (ICE)–powered vehicles. Furthermore, they are exclusively rear-wheel drive; Porsche has not announced any Carrera 4 variants as this book is written.

Base Carrera models continued with a revised Typ 992 3.0-liter twin-turbocharged flat six. By adopting the turbochargers from the previous GTS and the intercoolers from the Turbo, Porsche pulled another 9 horsepower from the engine, bringing total output to 394 horsepower. In a coupe weighing 3,505 pounds, this yielded a power-to-weight ratio of 1:8.9, in what many purists regard as the most enjoyable of all 911s.

Porsche engineers and designers continually reexamined existing systems. Even the folding cloth cabriolet top received an upgrade. It now opened or closed in 12 seconds at road speeds as high as 31 miles per hour. This required revising the three internal magnesium main elements while carrying over the heated fixed-glass rear window. For the base Carrera, options offered rear-axle steering PASM Sport Suspension, front-axle lifting (up to 1.6 inches), Sport Chrono package, sports exhaust, power steering plus, a 22.2-gallon fuel tank, and innumerable interior and exterior appearance and comfort options, including now-optional rear seats. Porsche recognized that few people ever rode in the rear seats and made them a no-cost option while saving weight and increasing storage capacity. Meanwhile, the exterior color palette has grown to 127 colors even before one sampled the rainbow of Paint to Sample.

But the elephant in the garage was the new GTS T-Hybrid. It should come as no surprise to anyone that the primary reason for this innovation was emissions. So it's most interesting that Porsche applies this still-controversial technology to the GTS lineup, which is surely the 911 GT3's nearest sibling in character and characteristics.

This is a Porsche *with* hybrid power; the 992.2 GTS T-Hybrid is not a front-axle add-on, nor is it a plug-in system meant to squeeze extra fuel economy out of the car in a silent and stealthy driving mode. In fact, there's no way to plug the car into anything but the traditional service station gasoline pumps. "T-Hybrid is the moniker Porsche has given to the performance hybrid power train," *Car and Driver* senior engineer Csaba Csere wrote on their website in late May 2024. "The T stands for Turbo and means performance. That, not fuel economy, is the focus of this new configuration."

To do this, engineers replaced the standard 2,891cc flat-six type 9A2B6. In a complex calculus, engine designers

SPECIFICATIONS

YEAR	2020-2024
DESIGNATION	992.1
MODEL AVAILABILITY	Carrera S, Carrera 4S, Carrera S Cabriolet, Carrera 4S Cabriolet; seven-speed manual; eight-speed PDK Porsche Torque Vectoring Plus (PTV-Plus), mechanical diff. lock for mt; electronic rear diff. lock for PDK
WHEELBASE	2,450mm/96.5 inches
LENGTH	4,519mm/177.9 inches
WIDTH	(w/o mirrors) 1,852mm/72.9 inches
HEIGHT	(S, 4S coupe) 1,302mm/51.3 inches (S, 4S cabriolet) 1,299mm/51.1 inches (S, 4S cabriolet) 1,299mm/51.1 inches
WEIGHT	(S coupe manual) 1,480kg/3,263 pounds (4S coupe manual) 1,530kg/3,373 pounds (S Cabriolet PDK) 1,585kg/3,494 pounds (4S Cabriolet PDK) 1,635kg/3,605 pounds
TRACK FRONT	1,589mm/62.6 inches
TRACK REAR	1,557mm/61.3 inches
WHEELS FRONT	8.5Jx20 ET 53
WHEELS REAR	11.5Jx21 ET 67
TIRES FRONT	245/35ZR20
TIRES REAR	305/30ZR21
CONSTRUCTION	Monocoque Steel/Aluminum/Magnesium
SUSPENSION FRONT	Spring-strut, wishbones with trailing links
SUSPENSION REAR	Multi-link, PASM
BRAKES	F/R 350.5mm/13.8 inches, ventilated, cross-drilled discs
ENGINE TYPE	Horizontally opposed water-cooled DOHC four-valve six-cylinder; twin-turbocharged and intercooled, direct fuel injection Typ #9A2B6
ENGINE DISPLACEMENT	2,981cc/182.0 CID
BORE AND STROKE	91.0 x 76.4 mm/3.58 x 3.01 inches
HORSEPOWER	443@6,500rpm
TORQUE	3,90lb-ft@2,300rpm
COMPRESSION	10.2 : 1
FUEL DELIVERY	Direct fuel injection, twin turbochargers, intercoolers
TOP SPEED	S Coupe manual 191mph 4S Coupe manual 190mph S Cabriolet PDK 190mph 4S Cabriolet PDK 189mph

determined the best means to meet coming European legislation related to Lambda emissions was by enlarging the engine, seemingly flying in the face of the early 992 policy of "Smart Sizing," that is, downsizing and turbocharging. (Lambda, represented by λ, refers to the air-fuel equivalence ratio; the ideal is 1:1, that is, an engine running neither too rich nor too lean. Engineers know there are times when running rich is beneficial to engine longevity, so Porsche relies on supplemental electric power to fill in power loss when going for that ideal, so called λ = 1.0.) Engineers enlarged cylinder bore and lengthened piston stroke from 91.0mm by 76.4mm up to 97.0mm by 81.0mm to yield 3,591cc. The redesign reduced overall height of the engine by 4.3 inches. The new engine, designated 9A3B6, needed to fit within the engine compartment of the 992.2 body while allowing room for the hybrid system's electronic power controls. The oil filter migrated from the top to the bottom of the engine to retain its easy service accessibility.

The highly developed VarioCamPlus valve operation system is gone, "which is no longer deemed necessary to broaden the torque," Csere wrote. The electric motor handled that duty—and others. "These heads also have larger intake and exhaust ports to accommodate greater throughput through the engine." These changes and others, including trimming the heft of the crankshaft counterweights, reduced the new engine's total weight by 40 pounds compared to the previous 9A2B6.

One of the benefits of twin turbos was faster spooling up turbine spin speeds to achieve full boost. With the new engine, they returned to a single turbo, the first time since 1994, but one with a pure racing innovation inside. Engineers fitted an electric motor inside the turbo housing, inserted between the exhaust-driven turbine and the compressor turbine. Its electric power kept the turbine spinning at 125,000 rpm even during closed throttle.

The single turbo required a new exhaust system with a crossover pipe from

▼ Porsche introduced the 992.2 GTS lineup with three models, the coupe, cabriolet, and this Carmine Red GTS Targa 4. Prices ranged from $166,895 for the coupe, to $180,195 for the cabriolet, and $187,995 for the Targa. Deliveries began at year-end 2024. *Porsche Presse*

▲ This is Porsche's ultimate 992, the 2025 Carrera, proving engineers still had more to develop on the car. Adopting GTS turbos and Turbo intercoolers boosted output by 9 horsepower. *Porsche Presse*

the left bank to the right where the turbo is mounted. As written above, this regenerative electric turbo is technology taken directly from Porsche motorsports applications, and in its latest use on the street, the motor-driven turbo reduces turbo lag at low engine speeds to nearly none. Csere again: "For example, at 2,000 rpm and a closed throttle, the old GTS required more than three seconds to achieve full boost [18.6 psi] when the throttle was floored, while the new GTS does it in slightly less than one second," and now achieved 26.1 psi of boost. Turbo lag is gone!

The ICE and turbo were not alone in redesign and reconfiguration. In some ways, it was the PDK that made the entire Hybrid drive possible. The redesigned eight-speed PDK split its gears on four shafts, not two. This shortened the PDK case to accommodate an electric motor within the case. But this rotating electric motor eliminated any possibility of developing a manual transmission for the GTS range. This big rotor also replaced the starter and generator. And by generating 400 volts, the AC needed a much smaller compressor than under a 12-volt system. What's more, engineers moved the water pump inside the engine, driving it off an extension of the oil pumps. This eliminated any external drive belt.

The 216-cell lithium-ion traction battery weighed 59.5 pounds and was roughly the same size as the standard 911 lead-acid battery; it fit in the same location, just forward of the windshield. What's more, the battery has its own liquid-cooling system to keep its temperature below 108 degrees Fahrenheit. A much smaller 12-volt battery (for those power-off uses) sits below the rear parcel shelf.

As if to prove the sincerity of "this Hybrid is for performance improvement only" claim, there is no clutch between the flat-six and the gearbox. There is no possibility of electric-only drive. Porsche 911 chief engineer Michael Rösler is quick to acknowledge their "goal was to make the car *not* seem like a traditional hybrid system with it usual stigma: 'It's more weight, it's less performance, it's not cool.' We definitely hope this is the world's first cool Hybrid," he admitted with a grin. The electric motor in the PDK added 54 horsepower to the 3.6-liter's 478, providing 532 horsepower at 7,500 rpm, and its 110 lb-ft of torque brought the total to 449 lb-ft of torque between 1,950 and 5,000 rpm. Porsche quoted the car's weight at 3,516 pounds, so with its combined horsepower, the car had a power-to-weight ratio of 1:6.61. While the car is 110 pounds heavier than its predecessor, this is primarily due to the Hybrid bits, which balance out this increase with additional horsepower. More surprisingly, its within pounds of the 992 Turbo with 12 horsepower more! Rösler acknowledged the new Hybrid GTS was every bit as suited—and intended—for track days as the 991 version had been. He stressed that the optional carbon-fiber roof, lightweight bucket seats, and lighter weight window glass will be available for those counting the ounces.

For model year 2025, Porsche offered the revised 911 Carrera starting at $122,095. Porsche priced the new 992.2 GTS T-Hybrid coupe at $166,895 and cabriolet at $180,195. The Carrera 4 GTS Targa T-Hybrid starts at $187,995. Base Carrera 992.2 deliveries started in late summer 2024, and the 2025 Carrera GTS range arrived in dealers at year-end.

PATIENCE WILL BE REWARDED

By August 1, 2024, enthusiasts were awaiting word of the 992.2 GT3 and GT3 Touring. A number of internet websites carried video clips of cars undergoing high-speed tests under body cladding and camouflage. Educated guesses predicted a slight horsepower increase coupled with extremely intense attention to drag and aerodynamics and on further improving the already fine roadholding and handling. The playing field for highest-performance GT cars has grown congested; Aston Martin, BMW, Chevrolet, Ford, and Mercedes-Benz have joined the GT3 racing grid with McLaren, Ferrari, and Lamborghini to unseat Porsche as the fastest thing around the track. In physical appearance terms, Flacht's biggest efforts were devoted to drag reduction and further optimizing aerodynamics. Another goal was further sharpening the suspension. One big question loomed over the new car. Observers have noticed that all the testing mules were equipped with PDK; no one has heard a test driver manually shift a gear. But if the recent uproar over the decision to limit GT3s to PDKs offered any hint of the future, those with suitable checkbooks have already formulated the ready-to-send email canceling their advance order if the car is PDK only. Recent comments were encouraging, with one GT Cars manager admitting they were looking at options—not only for the GT cars but also for the Carrera and possible Carrera S models.

And "possible" is all that's known. As Porsche entered the 2025 model year, it was carrying over some 15 previous generation 992 models—at least in U.S. markets—ranging from the Carrera T, the full run of Carrera S cars, all the Turbos, the GT3 RS, and the S/T. One presumed, or guessed, or hoped, that one's favorite from this list was on the list for updating.

A MARVEL. A FLAGSHIP. A TOUR DE FORCE

Is the new GTS T-Hybrid, available in 2025, a marvel? Or is it Porsche's new flagship, at least temporarily dislodging the long-standing halo over the head of the Turbos from that distinction? One thing is sure: the λ=1.0 target is industry wide across Europe and elsewhere, and as long as it remains an official emissions data point, Porsche's solutions will spread across its lineup. What Weissach hopes is that other carmakers carefully scrutinize this technology and determine Porsche's system is the one they must license as well to meet this critical goal.

So it's a marvel and a flagship and an engineering and electronics tour de force. It is also a great encouragement against those naysayers complaining vociferously that they already hate the future. Even as this is written, Porsche announced that by model year 2030, fully 80 percent of Porsche's vehicle lineup will be electric. Those who prefer their cups half empty will wail about that noiseless 80 percent while ignoring they're also talking about stylish vehicles with excellent roadholding and stratospheric fun-to-drive quotients that also are harming the environment less. Those who celebrate their cups half full will understand that cars such as the GTS T-Hybrid are what represent that other noisy 20 percent of Porsche's next-decade future: smart looks, spectacular handling, and greatly improved internal combustion engine performance that benefits from regenerated electric energy. And with the new Carrera, GTS, and coming GT3 models, there is still plenty of noise.

▲ This is the legendary "base" 911 Carrera, the simplest, least adorned model Porsche makes in the lineup. And to many enthusiasts and purists, this is also Porsche's best 911, with 394 horsepower in a 3,505-pound rear-drive coupe. *Porsche Presse*

INDEX

12 Hours of Sebring, 99, 203
24 Hours of Daytona, 327
24-Hours of Spa, Belgium, 190, 212

A

Achleitner, August, 255, 284, 285, 293–294, 339, 340
Aerokit II, 216
Aichele, Tobias, 33
Akin, Bob, 99
Alban, Manny, 354
Alusil, 82
Ampferer, Herbert, 196, 209, 210
Andersson, Åke, 60
Apenbrink, Boris, 345, 346, 352
Asch, Roland, 173
ATE disc brakes, 27, 55
ATS wheels, 76, 83
Audi Quattro, 138
Audi, 162, 190
AvD Oldtimer Grand Prix, 327

B

B series, 23, 54
Baillot-Lena, Claude, 149
Bantle, Manfred, 141, 142, 170
Barbour, Dick, 118, 119
Barnes, Tom, 52
Barth, Jürgen, 79, 107, 108, 138, 145, 149, 158, 174, 175, 178, 179, 181, 191, 202, 203, 210–212, 246, 248
Baur, Karl, 51
Beierbach, Walter, 41
Berger, Wolfgang, 72, 74, 75, 49, 94, 96
Bertone, Nuccio, 44
Bez, Ulrich, 170, 178, 195, 197, 199–201
Bezner, Fritz, 164, 165, 170, 174, 197
Bi-Xenon headlight modules, 244, 270
Bilstein shocks, 173
Biral cylinders, 74
Blaube, Wolfgang, 52
BMW, 72, 118, 129, 170, 190, 195, 251
Boge hydro-pneumatic front suspension, 55, 70
Boge shock absorbers, 146
Bohn, Arno, 191, 200
Bongers, Marc, 205
Borg-Warner, 150, 151
Bosch, 54, 77, 103, 142
    D-Jetronic, 92
    K-Jetronic fuel injection, 76, 79, 84, 87, 95, 115, 121, 122
    L-Jetronic, 128
    Motronic system, 149
    2.10 system, 202
    ME 7.2 engine management system, 227
Bose Surround Sound audio, 264
Bott, Helmuth, 14, 27, 40, 42, 43, 46, 47, 67, 71–73, 107, 138, 141, 142, 146, 158, 159, 161, 162, 164, 166, 170, 191, 192, 197
Böttinger, Astrid, 333
BPR series, 202, 203
Brands Hatch, 212
Branitzke, Heniz, 67, 165, 191
Brodbeck, Tilman, 66, 71, 73
Buchmann, Dieter, 285
Buchmann, Rainer, 93, 285

C

C Program, 122
C series, 23, 65
Car Top Systems (CTS), 258
Carrera Cup, 248, 250
Carrera Cup wheels, 182, 189
Carrera Power Kit, 248
Carrosserie, Reutter, 7, 9
Carrozzeria Pininfarina, 52
Chandler, Otis, 146
Chevrolet, 190
Corvair, 42, 43
Corvette, 47
Christiansen, Tomas, 258
Cup-Design wheels, 171

D

D Program, 122
D series, 65
D'Ieteren Feres, 14
Daytona, 50, 76, 149
Dick, William, 49, 51
Diedt, Emil, 52
DME Motronics computer, 128, 142
Donahue, Mark, 74
ducktail, 73, 74, 77, 82, 164, 252
Dunlop tires, 142, 149
DuPont Aramid fiberglass-reinforced epoxy resin, 145
Dupuy, Dominique, 191
Dürheimer, Wolfgang, 255

E

E Program, 127
E series, 71, 75
E-Gas, 227, 244
East Africa Safari, 93, 104
Ehra-Lessien test track, 73
Endurance Driver's Championship, 119
Engines
    901/02, 47
    911/93, 84
    911 Carrera, 362–363
    911 GT3 R, 335, 347
    911 RSR, 327–328
    911 Turbo, 343, 344
    1500 S Typ 528/2, 12
    1600 Normal, 16, 18
    Carrera, 74
    M30/69, 173
    M30/69SL, 175
    M64.70, 207
    M64/01 Cup, 173
    M64/04, 193
    M64/05, 196
    M64/21, 200
    M64/50, 184
    M64/60S, 217
    M64/81, 203
    M64/83, 206
    M64/C4L, 179
    M96/01, 221, 224
    M96/03, 245, 248
    M96/05, 256
    M96/79, 242
    M96/80, 212
    M97/76, 243
    M97/80
    Super 90, 13
    Typ 369, 11
    Typ 506, 10
    Typ 546, 12
    Typ 547, 67
    Typ 616, 40
    Typ 616/1, 14
    Typ 616/36, 56
    Typ 616/6, 28
    Typ 616/7, 24
    Typ 692/2, 20
    Typ 695, 13
    Typ 745, 17, 40
    Typ 753, 18
    Typ 821, 18, 22
    Typ 901, 18
    Typ 901/01, 31, 32, 45
    Typ 901/06, 55
    Typ 901/22, 61
    Typ 911/03, 66
    Typ 911/53, 72
    Typ 911/77, 105
    Typ 911/83, 77
    Typ 930/04, 103
    Typ 930/09, 115
    Typ 930/10, 121
    Typ 930/18, 125
    Typ 930/20, 131, 153
    Typ 930/21, 131
    Typ 930/25, 94, 150
    Typ 930/50, 88
    Typ 930/60, 111
    Typ 930/60S, 156
    Typ 930/61, 112
    Typ 930/72, 97
    Typ 930/73, 97
    Typ 959/50, 139, 140
    Typ 992 Carrera S/4S, 339
    Typ M64/01, 163
    Typ M96/77, 267

F

F Program, 145
Falk, Peter, 50, 74, 174, 195, 197
Farina, Sergio, 52
Fatthauer, Jon, 285
Fédération Internationale des Automobiles (FIA), 50, 72, 74, 76, 82, 93, 94, 96, 107–109, 138, 148, 174, 182, 202, 212, 234, 250, 267
Fère, Paul, 108, 112
Ferguson, Harry, 141
Ferrari, 129
Ferrari, Enzo, 107
Fields, Ted, 107
Fitzpatrick, John, 110, 111, 118, 119
Flegl, Helmut, 74, 178
Ford Motor Company, 15, 50, 72, 162
Frankel, Ingo, 282
Frankfurt International Auto Show, 23, 31, 40, 80, 88, 92, 138, 146, 158, 198, 200
French Grand Prix, 31
Frere, Paul, 212
Frankfurt International Auto Show, 332
Fuchs wheels, 46, 47, 52, 61, 76, 282
Fuhrmann, Ernst, 9, 17, 20, 52, 67, 68, 72, 74, 79, 96, 115, 116, 118, 120, 122
Fujiyama circuit, 50

G

G Program, 146
G series, 82, 84, 98
General Motors, 42, 56
Geneva International Motor Show, 44, 123, 205, 264, 265, 282, 326, 333, 334
Getrag transmission, 222
G50 transmission, 150, 151, 158, 196
Glemser, Dieter, 62
Gmünd, Austria, 7, 8, 40
Goertz, Albrecht, 12, 13, 44, 67
Goodwood Festival of Speed, 335
Gouhler, Joel, 191
Guilleaume, Paul von, 331

H

H Program, 150
H series, 88
Hatter, Tony, 170, 174, 195, 201, 209, 210
Hensler, Paul, 46, 65, 69, 79, 164, 165, 170, 195
Herschel, Ernst, 8
Hoffman, Max, 9, 12, 44, 47, 356
Holbert, Al, 146

I

Ickx, Jacky, 141
IMSA, 118, 148, 149, 187
IROC series, 93

J

J Program, 151
J series, 90
Jennings, Bruce, 46
Jensen, Allan and Richard, 141

K

K Program, 158, 166
K series, 92
Kahnau, Bernd, 197, 258
Karmann of Osnabruck, 14, 17, 18, 24, 40, 41
Kauhsen, Willi, 62
Kern, Lars, 329
Kiefer, Ekkehard, 148
Klie, Heinrich, 12–14, 27, 40, 331
Knirsch, Stefan, 221, 256
Komenda, Erwin, 7, 10, 11, 13, 32, 34, 42, 255
Koni shock absorbers, 47, 48
Konradsheim, Georg, 76
Kremer, Erwin, 110, 111, 124
Kremer, Manfred, 110, 111, 112, 124
Kulla, Matthias, 258, 295, 319
Kussmaul, Roland, 104, 142, 148, 173, 178, 203, 204, 235

L

L Program, 166
Lagaay, Harm, 195, 205, 222, 227, 257
Lai, Pinky, 222, 224

lambda sonde oxygen sensors, 115
Lapine, Anatole "Tony," 56, 68, 71, 73, 74, 76, 84, 123, 131, 170
Larson, Grant, 309, 310, 345, 352

Le Mans, 31, 79, 97, 107, 108, 110, 119, 140, 141, 148, 149, 190, 191, 202, 206, 207, 209–211, 234, 347
Lemoyne, Dominique, 141
Lincoln Continental, 15
Linge, Herbert, 62, 63
Loos, Georg, 79, 110
Los Angeles Auto Show, 264
Lotz, Kurt, 56
Lowney, Damon, 354, 355
Ludvigsen, Karl, 69, 112
Ludwig, Klaus, 112
Ludwigsburg, 92, 103
Lutz, Bob, 118, 120

M

M Program, 179
MacPherson struts, 14, 226, 231, 349
Marathon de la Route, 63, 325
Marchart, Horst, 170, 197, 222
Maurer, Michael, 340
McLaren, 212
MECOSA (mechanically converted shift actuator), 341–342
Mercedes-Benz, 43, 73, 118, 129, 141, 212, 251, 257
Metge, Rene, 141, 142, 149
Mezger, Hans, 17, 18, 23, 32, 50, 74, 76, 79, 107, 108, 142, 148, 234, 243, 244, 248, 259, 264, 265, 274, 283
Michelin Supercup, 248, 265
Michelotti, Giovanni, 43
Mickl, Josef, 14
Minilite wheels, 61
Möbius, Wolfgang, 84, 112, 162
Models
356, 10, 22, 23, 27, 34, 42, 46, 56, 74, 76, 118
356/2, 8
356–1600C, SC, 40
356A, 7, 10, 14, 15
356B, 13, 28
356C, 26–28
356 Super C, 27, 28
530, 13
695 prototype, 44
718 Formula 2, 12
804, 40
901, 22, 23, 27, 31, 34, 37, 38, 122
901, 1964, 31–39
902, 27, 31, 40
904, 40
904 Carrera GTS, 12
906 Carrera 6, 61
911 Carrera, 1984–1986, 127–137
911 Carrera, 1987–1989, 150–156
911 Carrera, 1989–1993, 161–169
911 Carrera, 1994–1997, 195–201
911 Carrera, 1998–2001, 221–233
911 Carrera 2025, 361–364
911 Carrera GTS 2011–2012, 287–288
911 Carrera GTS/4 GTS 2018–2020, 329
911 Carrera S, 1997–1998, 216
911 Carrera T 2018, 332–333
911 Carrera, Turbo, 2002–2004, 244, 245
911 Dakar, 356–358
911 GT1, 1996–1997, 209–215
911 GT2, 2003–2005, 246, 247
911 GT2, GT2 Evo, 1995–1997, 202–208
911 GT2 RS 2010, 288–289
911 GT3 RS 2023, 353–355
911 GT2 RS Clubsport, 336
911 GT3, 1999–2005, 234–243
911 GT3 2018, 329
911 GT3 2022, 346–348
911 GT3 R 2019, 335
911 GT3 RS, 2012, 290–291
911 GT3 Touring 2018, 332
911 GT3 Touring 2022, 346–348
911 GTS 2017–2018, 328–329
911 GTS 2022, 348–349
911 GTS 2025, 361–364
911 GTS Cabriolet America Edition, 355–356
911 Porsche Design Targa 4S, 355–356
911 RSR 2017, 327–328
911 SC 3.1, 114
911 SC/RS, 145
911 Speedster, 158, 286, 289
911 Speedster Concept 2019, 335
911 Sport Classic, 352–353
911 S/T, 358–361
911 T 2023, 349–352
911 Targa, 2001–2005, 248, 249
911 Targa 4/4S 2021, 345–346
911 Targa 4S 2016, 326–327
911 Targa, Carrera, 2007–2010, 268–273
911 Targa 50 Years Porsche Design, 349
911 Turbo, 1978–1979, 111–113
911 Turbo, 1980–1985, 114–125
911 Turbo, 1991–1992, 170–177
911 Turbo, 2008–2009, 278–281
911 Turbo 2018, 334
911 Turbo 2021, 343–345
911 Turbo S, 1997–1998, 216–219
911 Turbo S 2021, 343–345
911 Turbo S Exclusive Series 2018, 331–332
911 Typ 991 Carrera, 293–297
911 Typ 991 GT3, 2013, 301–302
911 Typ 991 Targa 4, 2015, 309–313
911 Typ 991 GT3 RS, 314–315
911 Typ 996 Turbo, 2014–2015, 302–309
911 Typ 991/2 Carrera, 2017, 315–319
911 Typ 991/2 Carrera 4, 2017, 319
911 Typ 991/2 Turbo, 2017, 319–322
911 Typ 992 Carrera S/4S 2020, 339–343
911 Typ 992.2 GT3/GT3 Touring, 365
911, 1964–1967, 40–45
911, 1974–1975, 82–87
911/B17, 52
911/C20, 52
911E, 70
911GT, 50
911L, 1968, 54, 55
911R, 51, 52, 75
911R, 1967, 50–53
911R, 2017, 322–326
911S, 1967–1968, 46–49
911S, 1972–1973, 71–74
911S, 47, 48, 51, 52, 54, 75, 88
911SC, 122
911SC, 1978–1979, 103–106
911SC, 1980–1983, 114
911SC, 1983, 129
911ST, 50
911T, 1968, 61
911T, 1970–1971, 65–70
911T, 48, 50, 54, 79
911T/R, 50
911.2 2019, 334
912, 33, 40, 44, 45, 48, 54
912 E, 1976, 92–94
912, 1965–1969, 56, 57
914/6, 79
917 Can-Am, 80
917 InerSeries, 80
917–30 Can-Am, 82
922 Carrera 40th anniversary edition, 2003–2004, 250, 251
924, 92, 96, 98, 115, 118, 120
928, 115, 118, 120, 166
930, 115
930S, 150, 157
930 Turbo, 103, 118
933 Carrera, 200, 201
934, 94
934, 1976–1977, 94–100
934K, 111
935–78 Moby Dick, 1978, 107–110
935, 1976–1977, 94–101
935/77 racer, 1977, 99
935K, 118, 124
935K3, 1979, 111, 112
944 Turbo, 166
959, 158, 159, 162, 190
959 Komfort, 142, 147
964, 166, 167, 191, 195–197, 199–201, 222, 224
964 C4, 174, 197
964 Carrera Cup, 178, 179
964 RS, 178
964RS, 181, 183, 187
964 RSR, 174
965 Turbo, 196
993, 222, 224, 225
993 Carrera Cup, 201
993 Supercup, 203
993 Turbo, 205
996, 221, 223–225, 229, 247, 248
997, 256, 257, 258, 274
997/2, 268
997 GT3 RSR, 266
997 Speedster, 289, 290
997 Twin Turbo, 264
America Roadster, 9, 181
Baby 935, 96, 107
Can-Am, 79
Can-Am 917–30, 74
Carrera, 9, 43, 73, 82, 83, 86, 88
Carrera, 2005–2008, 255–263
Carrera, 2013, 297–300
Carrera 2 (C2), 17, 27, 165–167, 174, 184, 191, 195
Carrera 2.8 RSR, 79–81
Carrera 3.0 RSR, 1974, 82
Carrera 3.0, 90, 93
Carrera 3.2, 127
Carrera 3.2 Speedster, 154
Carrera 4 (C4), 165–168, 178, 179, 191, 195
Carrera 4 Lightweight, 1990–1991, 178, 179, 181
Carrera 4S, 206
Carrera 6, 50
Carrera 904GTS, 31
Carrera Cup C2, 181
Carrera GT, 13, 20
Carrera RS, 1973, 75–78
Carrera RS, 92
Carrera RS, RSR 3.8, 1993, 190–193
Carrera RSR 3.0, 93
Carrera T 2018–2020, 333
Carrera Turbo Look, 1992–1994, 181–189
Centurion Edition, 2010, 284
Club Sport Carrera, 151–153
Continental, 15
Convertible D, 13
CTR 001, 158
CTR II, 206
Gruppe B Evolution rally, 124
GT2, 2008–2009, 274–277
GT3, 2006–2008, 243, 264–267
GT3, 2010, 282, 283
GT3, 2013, 301
GTS 4, 2015
InterSeries, 79
K3, 119
Moby Dick, 138, 142, 148
RSR Turbo, 93
RUF CTR Yellow Bird, 1987, 158, 159, 206
S, 1972, 72
Super 90, 24
T-6, 14, 26, 28
T-7, 27, 28
T-8, 14
T/R, 60, 61
Turbo, 1975–1977, 88–91
Turbo, 2009–2010, 284–286
Turbo Look Cabriolet, 181, 182
Turbo Targa, 93
Typ 356 B (T5), 326
Typ 745 T7, 41
Typ 804 Formula One racer, 12
Typ 901 No. 057 1964, 333
Typ 906 Carrera 6, 40
Typ 911 GT3 RS 2018, 333–334
Typ 933, 198
Typ 935 Clubsport 2019, 335–336
Typ 959, 1986–1988, 138–149
Typ 960, 141
Typ 961, 141, 148

Typ 964, 162–165
Typ 989, 195, 197
Typ 992.2, 361–364, 365
Typ 993, 199, 216
Monte Carlo Rally, 60, 68, 105
Monza, 50, 52, 325
Moretti, Gianpiero, 110
Morse, Kerry, 175
Mössle, Erhard, 258, 303–305, 307, 312
Murkett, Steve, 167

N

N Program, 183
Nader, Ralph, 42
Nardo circuit, 158
Nierop, Kees, 149
Nikasil, 74, 82
Niko, Michael, 269
Nordhoff, Heinz, 52, 56
Norisring, 110
Nürburgring, 63, 110, 174, 190, 327, 329

O

Onger, Kendrick, 142
Opel, 56, 84
Options
    Club Sport package, 334
    M030, 182
    M471, 75, 76
    M472, 75, 76, 123
    M473, 120
    M491, 75
    M503, 189
    M504 RSA, 182
    M505, 123, 130
    M505/506, 150
    M506, 146
    M701, 114
    Night Vision, 339
    Performance Kit, 124, 156, 187, 216, 218, 273
    Turbo Look, 131, 145, 154, 155, 158
    Weissach package, 334
    X50, 247, 256
    X51 Power Kit, 268

P

P Program, 183
Paris Auto Salon, 31, 40, 54, 76, 335
Paris–Dakar Rally, 141, 149, 174
PDK manumatic transmission. See Porsche Doppelkupplungsgetriebe (PDK) transmission.
Pelly, Chuck, 52
Penske, Roger, 82
Peter, Patrick, 202
Petersen Museum, 37
Peugeot, 31, 40
Piëch, Ferdinand, 32, 40, 50, 52, 56, 57, 66–69, 138, 325
Piëch, Karl, 23
Piëch, Michael, 67
Pirelli racing slicks, 234

Pirelli Supercup, 234
Pirelli tires, 91, 234
Porsche Active Drivetrain Mount (PADM), 278
Porsche Active Suspension Management (PASM), 256, 265, 268, 328, 332, 343, 345, 347, 350, 353, 356, 357, 362
Porsche Carbon Composite Brakes, 219
Porsche Cars North America (PCNA), 146, 175, 176, 182, 244
Porsche Ceramic Composite Brakes (PCCBs), 244, 246, 268, 282
Porsche Club of America, 262
Porsche Communication Management (PCM) system, 270, 280
Porsche Design, 349
Porsche Doppelkupplungsgetriebe (PDK) transmission, 190, 269, 270, 271, 274, 278, 284, 286, 341, 342, 343, 347
Porsche Double-Clutch transmission, 270
Porsche Dynamic Chassis Control (PDCC), 332
Porsche Exclusive Manufaktur facility, 331, 335
Porsche Experience Center Carson, 327
Porsche Experience Center Hockenheim, 336
Porsche Experience Center Los Angeles, 327, 339
Porsche Experience Center Shanghai, 334–335
Porsche Launch Assist, 264
Porsche Stability Management, 227
Porsche, F. A., 9, 10, 13, 14, 18, 22, 23, 26, 27, 40–42, 47, 54, 56, 66, 67, 122, 131, 256, 284, 331, 349
Porsche, Ferdinand Anton Ernst, 7
Porsche, Ferdinand, 9
Porsche, Ferry, 7, 9, 11, 13, 14, 17, 26, 27, 33, 34, 38, 40, 41, 44, 56, 66, 67, 79, 116, 118, 180, 123, 151, 248, 255, 259, 285, 331, 349
Porsche, Louise, 67
Porsche Museum Special Exhibition, 333
Porsche, Peter, 67
Poulain, Herve, 97
Preston, Vic, 105
Preuninger, Andreas, 264, 283, 289, 329, 301, 353, 355, 360, 361
Project 530, 10, 11
Project 695, 12
Project A, 274

R

R Program, 183
Rabe, Karl, 12
Raether, 50
Raether, Wolfgang, 40

Rahal, Bobby, 99
Rally Kit 47
Ratel, Stéfane, 202
Recaro seats, 46, 47, 174, 190
Reimspiess, Franz, 10, 17
Reisinger, Peter, 84
Reitter, 212
Reitter, Horst, 209, 210
Rennsport Reunion, 335–336
Reutter body works, 17, 40, 41, 331
Richter, Les, 82
Riverside Raceway, 83
Ruf, Alois, 158, 206, 210, 211, 213, 229, 232, 233, 243, 246, 252, 253, 274, 285

S

Safari Rally, 104
Sauter, Uli, 257
SC series, 122
Schäffer, Valentin, 79
Schätzle, Michael, 295
Schröder, Gerhard, 27, 40–42, 122, 123, 331
Schulz, Eberhard, 93
Schulzki, Markus, 310–311
Schutz, Peter, 120, 122, 127, 138, 141, 146, 151, 158, 159, 161, 165, 166, 170, 191, 209
Shanghai International Circuit, 334
Siegert, Franz-Josef, 257
Silverstone, 110
Singer, Norbert, 66, 72, 74–76, 79, 82, 93–96, 107–109, 138, 142, 174, 109–212
Smale, Glen, 148
Soderberg, Dick, 84, 142, 143, 148, 162
Soeding, Otto, 32
Solex carburetors, 23, 32, 45, 46
Soukup, Emil, 43
Spartan configuration, 75
Sport Chrono Plus, 260, 264
Sport Kit I, 48
Sport Seats, 132
Sportomatic Transmission, 49–52, 63, 115
Sports Chrono Plus, 284
Sprenger, Rolf, 183
Steckkönig, Günther, 73, 148
Stewart, Rod, 100
Stout, Pete, 279
Strähle, Paul Ernst, 74, 173
Stuttgart, 9, 10, 52, 115, 118, 127

T

T Program, 205
T-5 body, 22
Teldix ABS system, 74
Tiptronic transmission, 168, 181, 183, 190, 191, 196, 205, 206, 222, 227, 233, 244, 251, 256, 259, 264, 279
Tomala, Hans, 17
Tour de Corse, 76
Tour de France, 202, 325
Toyota, 199
Triumph TR4, 43

Troutman and Barnes, 49, 52
Troutman, Dick, 52
Turin, Italy, 44

V

Vallelunga, 110
van Hulten, Ivo, 345
VarioCam Plus, 244, 246, 247, 256, 268
VarioCam system, 221, 244
VarioRam induction, 200, 205, 222, 234
VLN race series, 282
Volkswagen, 33, 42, 51, 52, 56, 66–68, 73, 92
von Hanstein, Baron Huschke, 40, 325
von Neumann, John, 9, 12, 42–44, 356

W

Wagner, Harlad, 40, 44
Waldegaard, Björn, 68 93, 105
Walliser, Frank-Steffen, 327–328, 336
Wasserbäch, Thomas, 294
Watkins Glen, 76
Weber carb, 46, 47, 54, 61, 70
Weinsberg body works, 40
Weissach, 52, 68, 73, 76, 79, 80, 88, 103, 107, 114–116, 145, 146, 161, 164, 165, 174, 190, 191, 195, 197, 202, 206, 210, 212, 227, 231, 234, 237, 240, 246, 248, 251, 255, 256, 259, 265, 274, 278, 285
Werke I, 76
Werke V, 161
Westinghouse anti-lock braking system, 142
Wet Mode, 339
whale tail, 88, 114, 116, 182, 183
Wiedeking, Wendelin, 161, 166, 192, 199, 200, 201, 205, 212
Wilhelm, Anke, 257
wind tunnel, 71, 72, 161
Winkelhock, Markus, 265
Wollek, Bob, 111
Woods, Roy, 99
World Endurance Championship, 118
World Endurance Championship Safety Cars, 118, 334
Wütherich, Rolf, 50

Zasada, Sobieslav, 93
Zell am See, Austria, 67
Zenith-Solex carb, 69
Zimmerman, Ekkehard, 100, 111, 123, 124, 285
Zolder, 111
Zuffenhausen plant, 27, 56, 65, 66, 75, 76, 82, 120, 146, 151, 183, 190, 191, 197, 199, 200, 205, 206, 225, 258, 285, 326
Zuhaï, 212